Unleash the System On Chip using FPGAs and Handel C

Unleash the System On Chip using FPGAs and Handel C

Rajanish K. Kamat • Santhosh A. Shinde •
Vinod G. Shelake
Authors

Department of Electronics
Shivaji University, Kolhapur
India

 Springer

Authors
Dr. Rajanish K. Kamat
Department Electronics
Shivaji University
Kolhapur-416004
India
E-mail: rkk_eln@unishivaji.ac.in

Mr. Vinod G. Shelake
Department Electronics
Shivaji University
Kolhapur-416004
India

Mr. Santhosh A. Shinde
Department Electronics
Shivaji University
Kolhapur-416004
India

ISBN: 978-1-4020-9361-6   e-ISBN: 978-1-4020-9362-3

Library of Congress Control Number: 2008942203

Printed on acid-free paper

9 8 7 6 5 4 3 2 1

springer.com

Preface

With the rapid advances in technology, the conventional academic and research departments of Electronics engineering, Electrical Engineering, Computer Science, Instrumentation Engineering over the globe are forced to come together and update their curriculum with few common interdisciplinary courses in order to come out with the engineers and researchers with muli-dimensional capabilities. The growing perception of the 'Hardware becoming Soft' and 'Software becoming Hard' with the emergence of the FPGAs has made its impact on both the hardware and software professionals to change their mindset of working in narrow domains. An interdisciplinary field where 'Hardware meets the Software' for undertaking seemingly unfeasible tasks is System on Chip (SoC) which has become the basic platform of modern electronic appliances. If it wasn't for SoCs, we wouldn't be driving our car with foresight of the traffic congestion before hand using GPS. Without the omnipresence of the SoCs in our every walks of life, the society is wouldn't have evidenced the rich benefits of the convergence of the technologies such as audio, video, mobile, IPTV just to name a few. The growing expectations of the consumers have placed the field of SoC design at the heart of at variance trends. On one hand there are challenges owing to design complexities with the emergence of the new processors, RTOS, software protocol stacks, buses, while the brutal forces of deep submicron effects such as crosstalk, electromigration, timing closures are challenging the design metrics. Moreover the time to market pressure, process roadmap acceleration, mixed design flows have posed various challenges to the SoC designer community. The present book 'Unleash the System On Chip using FPGAs and Handel C', attempts to address few challenging issues of SoC design leveraging the FPGA platform with 'C' based programming methodology.

Organization of the book

The book is divided into seven chapters. Chapter 1 introduces the concept of SoC. It visualizes the unique place of the SoC in the microelectronics arena and covers the market trends, challenges and opportunities for the budding professionals in this

field. Chapter 2 covers the essential details of the Handel C, which is extensively
used in the book for making software hard. The third chapter presents details of
sequential circuit design with appropriate case studies. The fourth chapter details
the combinational system designing from the SoC viewpoint on the Xilinx FPGA
platform. Chapter five throws light on placing the algorithms in the SoC paradigm
using Xilinx EDK platform with emphasis on 'Hardware-Software Codesign'.
Chapter six is on the theme of rapid prototyping with unique design realizations
of fuzzy logic controller, Network on Chip and ciphers. Chapter seven extends the
same theme, however with a different approach of using soft processor cores.

The book is carved for all those who wish to be SoC designers with relatively less
steep learning curve. One of the striking features is one can possibly realize the designs
presented here with simple setup and free open source tools such as system C.

In these economically turbulent times, escalating efficiency and dropping risk while
continuing to innovate would be decisive for the SoC professionals and industries
to ride out the economic slowdown. We are sure that this book will definitely serve
as 'Light House' to guide the SoC designer community.

Interacting through blog and website

A blog is created by the authors exclusively to interact with the readers of this book.
The URL for the blog is: http://drkamat.wordpress.com/ . We are looking forward
towards your posts, comments and design queries on this blog. The readers may as
well see the updates and the upcoming SoC cores at the Author's homepage at URL:
http://www.rkkamat.in

Foreword

System-on-chip technology (SoC) is a truly innovative and creative realm of VLSI that brings multiple functions integrated on a single silicon chip. According to the electronics.ca research network, the SoC Technology market is currently estimated at nearly $14.4 billion. The prophecy is further growth, with an average annual growth rate of 24.6%, reaching around $43.2 billion by 2009 i.e. by next year. The market statistics mentioned above gives an insight as regards to this important and ever growing segment of VLSI technology. The SoC with phenomenal growth rate is experiencing many challenges from the designer point of view such as narrowing development cycle without compromising the product functionality, performance, reliability and quality. The computational complexity makes the SoCs to stand apart in the crowd of the custom and semicustom systems and poses unique challenges to the designers that can be described as "More than Moore" or "greater than equal to Moore". While the major players in this field like Infineon, Motorola, Intel, Wipro are achieving an annual growth rate of more than 100%, still the industry is evidencing "More is Less" with ubiquitous penetration of SoCs in every walk of society.

How does the academic, research and industry keep up with the growing challenges of SoCs? Here is a book that rightly focuses on the issues pertaining to digital SoCs. It is a sort of brainstorming at the top level and then percolating the same at the individual modules with top-down approach. This is the challenge that Dr. Kamat and his research team faced when they were exploring various aspects of this field. Therefore, they have come out with a book that presents their hands-on experiences with digital SoCs. The book in its unique fashion addresses the design issues, technological challenges, and market legacies by presenting a distinctive know-how based on FPGAs and C based methodology.

The approach presented in the book is multifaceted. It is solidly grounded on a tripod approach comprising of Xilinx FPGAs, Handel C and soft IP cores. Dr. Kamat weds the trio with practical case studies and came out with a good number of building blocks to build successful design. It is vital to appreciate that the digital semi-custom design is not a straightforward process, especially in the backdrop of increasing software on the chip which is referred to as 'Software Parkinson'. The authors have successfully showcased the hardware software portioning and conveyed their notion to the readers.

Chapter 1 of the book presents various essential attributes of the SoCs. Extensive survey is conducted by the authors to present the state of the art systems, the design stats quo and various design and market challenges. Chapter 2 covers various aspects related to the C based methodology. The screenshots presented here makes the learning curve less steep for the novice designers. Chapter 3 and 4 presents various implementations of the sequential and combinational logic designs respectively. Chapter 5 gives an insight of customized arithmetic core designs. The striking feature of this chapter is coverage of Xilinx EDK suite, the no-fee intellectual property (IP) cores for designing embedded processing systems with Xilinx platform FPGAs. Chapter 6 has addressed the fuzzy logic controller and Network on Chip designs with which the authors have tried to inculcate intelligence on chip. This is further extended in Chapter 7 with hardware-software codesign methodologies using soft processor cores to realize the SoC design in narrowing time window.

Thus the authors have endeavored to present more than sufficient details to encourage the potential readers to dig at depth when dealing with the design principles that appears to be institutively obvious. Undoubtedly the design community could use the book to discover the programming concepts that actually aids in realizing the SoCs. They can as well extend the design methodologies presented in this book with the Open System C initiatives (OSCI) with the recent transaction-level modeling standard, TLM-2.0 that enables model interoperability and reuse, providing an essential framework for architecture analysis, software development and performance analysis, and hardware verification. Thus in lieu of the Handel C, the designers can apply TLM-2.0 (which is freely downloadable) to show their SoCs the light of the day.

To sum up I would like to quote that " The only constant in life is change" which is experienced in its real sense in the field of SoCs through Moore's law in Moore's own words, "Several times along the way, I thought we reached the end of the line, things tapered off, and our creative engineers come up with ways around them." The designer community will definitely appreciate that the present book is a piece of creative engineering.

I wish all the best to the potential readers to realize their SoCs with the know-how presented in the book.

Yogindra S Abhyankar

 **Yogindra S. Abhyankar** is a Group coordinator working with the hardware technology development group, Center for Development of Advanced Computing (C-DAC), Pune, India for more than 14 years. He has worked on various projects in high performance computing (HPC) involving high speed inter-connects, system architecture and VLSI design. His current field of research is in the area of "Reconfigurable Computing", one of the emerging approaches for speeding-up HPC applications. He holds a MS degree from the University of Wisconsin, USA.

Author's Profile

Dr. Rajanish K. Kamat is working as a Reader in the Department of Electronics, Shivaji University, Kolhapur, India. He enjoys teaching to Masters in Electronics and Masters in Technology classes. He is supervising number of research students working towards their doctorate in the area of VLSI Design. The uniqueness of his research work is its application orientation achieved through the analytical marriage of interdisciplinary themes such as neural network, Computational Techniques, Software etc. He has published his research work widely in the areas of Sensor Systems, VLSI Design, Embedded Systems, Network Security and Visualization techniques. He is recipient of Research Grants under the 'Young Scientist Scheme' from the Department of Science and Technology of Government of India. He is referee of good number of reputed research journals.

Mr. Santosh A. Shinde is working as a Research Fellow under the University Grants Commission scheme at Department of Electronics, Shivaji University, Kolhapur. He is also pursuing his Doctorate in the research area of VLSI Design which is in the final phase. His thesis topic is "Programmable ASIC Design for Circumventing SPAM". He also carries with him rich industrial experience in various branches of VLSI Design and Embedded Systems. He has published good number of research papers in referred International Journal of repute.

Mr. Vinod G. Shelake is working as a Lecturer at Sanjivani College, Panahala. He is also pursuing his Doctorate in the area of VLSI Design. His thesis topic is "FPGA based Firewall Design".

Acknowledgements

Many people motivated us to make this book possible. At the outset authors would like to acknowledge the grants received under DST-SERC Fast Track Project for Young Scientist SR/FTP/ETA-14/2006 entitled "Development of FPGA based open source soft IP cores for parameterized microcontroller design" to Dr. R.K. Kamat under which a grand repository of free SoC cores on web is coming up. Interested readers may keep a track on the web URL: http://www.rkkamat.in

Dr. Kamat owes great thanks to his teacher Dr. G.M. Naik for his inspiring support. Thanks are also due to Mr. Jivan Parab who will be coauthor in forthcoming interesting books on VLSI design. Thanks also go to authorities of Shivaji University, Kolhapur for the support received towards the infrastructure. Special thanks to Mr. Yogindra Abhyankar, CDAC Pune for reviewing the book and agreeing to give the foreword.

Dr, Kamat would also like to thank his wife Rucha, parents, students, friends and supporters for their support. Again special thanks to Adeet and Pari Shanbhag for their enjoyable company that relieved the stress of SoC designing.

Finally, special thanks to Mr. Mark de Jongh, Senior Publishing Editor and Mrs. Cindy Zitter from Springer for their outstanding support all the time. It is only because of them we feel like contributing quality work to the designer community through the reputed network of 'Springer'.

- Dr. R.K. Kamat
- Mr. Santosh A. Shinde
- Mr. Vinod G Shelake

Contents

List of Figures

List of Tables

List of Programs

Chapter 1
Introduction

The electronics industry is an amazingly fast changing and vigorous environment, driven by everlasting demand for innovation and technological advancement. As electronic manufacturers need to develop affluent features that will set apart new products, devices attain progressively more functions and head towards the complexity beating the Moor's law. These have resulted in narrow market windows and accelerate obsolescence cycles that naturally impose reduced development times.

State of art electronic devices requires chips that dwell in a small footprint, less hungry interms of power, and fast enough to deal with the sophisticated applications. Traditionally, semiconductor devices were developed with a notion of single, dedicated function. A system used to be analytical marriage of the subsystems for the intended application. However, the new era is all about the complete systems on a single chip. The state of the art systems have multiple processors or similar complex functional blocks interacting with each-other and with the real world as well to satisfy the needs of the appliance. In recent times SoC has emerged as the promising solution to all the above mentioned requirements. Owing to their applicability in the sophisticated systems, the SoC's are finding their place mainly in computers, communication equipment, consumer electronic devices, and automotive applications. The Soc Market is fuelled by their growing demand in new consumer applications, which include mobile phones, and automotives. Yet another factor that has escalated the SoC demand is the growing digital convergence and the technological progress towards the heterogeneous systems being placed on-chip. However the end result faced by the designer community is tremendous pressure with the shrinking time-to-market window. The complexity and heterogeneity of today's SOCs has forced the designers to devote more time for the simulation, verification and prototyping aspects.

Like all electronic systems, System-on-Chip (SoC) are subject to the contradictory trade-offs such as increase in complexity Vs decrease in time and costs. In addition, specific SoC characteristics aggravate difficulties in their design with growing multidisciplinary applications, closer hardware software codesign issues, demand for small footprint, custom Vs semicustom platforms to name a few.

R.K. Kamat et al., *Unleash the System on Chip using FPGAs and Handel C,*
DOI 10.1007/978-1-4020-9362-3_1, © Springer Science+Business Media B.V. 2009

Chapter 1 will be introducing the leverage on an extensive intellectual property (IP) portfolio, systems knowledge and experience with technologies like ARM to target specific vertical markets with System-on-Chip solutions that meet a superset of application needs.

1.1 Prologue

System-on-Chip (SoC), the new brawny arm of the VLSI is today's hot buzzword. "VLSI" technology is being evolved since its inception in an exponential way and more so with its convergence with other developing technologies such as telecommunication, consumer electronics, defense electronics to name a few. This simultaneous progress of the above mentioned technologies has thoroughly influenced the VLSI paradigm. 'Design Automation' or 'Computer Aided Design' that was in a very primitive stage in 70's is well matured today with the incorporation of the heuristics and intelligent algorithms typically running on a workstation to yield useful revisions in the design inputted.

It is worthwhile to estimate the classification parameters so as to get a clear cut idea of the design volume of today's VLSI chips and to locate where the SoC fits on the VLSI canvass.

In the literature the word ULSI i.e. Ultra Large Scale integration is also used interchangeably with VLSI, however more precisely it refers to the number of component count higher than 10^7. The technical fraternity is also divided over the issue whether to take into the account the number of gates or transistors to demark the scale of integration. It is therefore more appropriate to use number of components as the basis as shown in Table 1.1. The SoC conception in the VLSI arena has emerged out of the need of the state of the art products dedicated towards a single functionality.

1.2 Exceptional Attributes of the SoC Technology

VLSI technology has evolved from first planar integrated circuit having two transistors in 1961 to a Pentium 4 processor having 42 million transistors in 2001. This exciting and knowledgeable sector of Electronics industry has evidenced a 53% compound annual growth rate over 45 years. It is noticeable because no other technology has grown so fast so long. One of the factor that has fuelled this growth is the benchmarking standards provided by the visionaries such as Gorden Moore. Starting from the pure Electronics the VLSI has now penetrated in the lives of a common man; thanks to the growing interdisciplinary approach with the participation of the Electrical, Electronics, Biological, Artificial Intelligence professionals coming together to make the chips intelligent, fault tolerant and adaptive to the ambient environment. The interdisciplinary developments mentioned above have fuelled the development of the SoCs.

Table 1.1 Evolution of VLSI technology

Year	Technology	Maximum number of components / chip	Design methodology
1959–1965	Small-scale integration (SSI)	10^2	Manual (translation from electronics to graphics)
1965–1969	Medium-scale integration (MSI)	10^2–10^3	Manual but transition towards design automation
1969–1989	Large-scale integration (LSI)	10^3–10^5	Design abstraction and use of languages such as VERILOG and VHDL started
1989–till date	Very large-scale integration (VLSI)	10^5–10^7	Sophisticated compute intensive EDA tools such as simulated annealing, floorplanning, critical path analysis, time skew, power optimization
1990 onwards	Ultra large scale integration ULSI	Chips more than 1 million transistors	
1990–till date	System-on-Chip	Million gates + customized hardware + software	
1990 onwards	Wafer-scale integration		Entire silicon wafer is used to produce a single 'super-chip'.
	Three dimensional integrated circuit	Multilayer active electronic components	On-die signaling using vertical interconnects, wafer-on-wafer, wafer-on-die, die–on-wafer

   The main attributes of today's SoC systems is their complexity, small footprint, spatial and temporal efficiency and ability to design with a small design team empowered by the sophisticated computational tools known a EDA tools. SoC design industry has shown the world few innovative things and set up certain innovations in the overall design methodology. "Design and Reuse" is one such principle devised by which saves the efforts in reinventing the wheel and empowers the designer with a rich and well tested library of the IP cores with well defined I/O interfaces. This principle was laid down way back in late 1970s by Carver Mead and Lynn Conway whose textbook showed how to use library elements with proven behavior and design rules to create "cookbook" designs.

Integration of third party tools in the design flow is yet another innovation the SoC industry has achieved by agreeing upon the standards and formats of the output of each design step. This offers the freedom for the designer to make use of the best tool to optimize the design. Mixed Design flow is now a common practice in the SoC field.

Specialty of the SoC design paradigm is to do the same things in alternate number of ways. A mere working design is not the final goal, but the same can be improved by employing novel techniques. For instance the architecture of any design can be written by employing the behavioral, data flow or structural architectures. Today the SoC industry is trying to get the synergisms of the hardware and software professionals by stressing upon the abstract design flow by offering the designers C, C++ based environment that has been explored in-depth in the present book. This not only makes the design teams participative but also helps in making the design efficient as the well proven software algorithms can be utilized.

Design is easier in SoC than in other industries (as will be evidenced in the present book) such as mechanical systems because, the information at the system level is entirely logical and connective. This information is transformed and augmented from stage to stage in the design process but its essential logical/connective identity is preserved all the way to the masks. This is not possible in mechanical systems, where the abstractions are not logical homologues (much less homomorphs) of the embodiments and likely never will be. Instead, tremendous conversion is needed, with enormous additional information required at each stage [23].

An enormous variety of SoC products can be obtained with varying specifications, though the flow is common due to the different manners of specifying the system.

1.3   Classical Taxonomy: A Holistic Perspective Extended Towards Integrated Circuits Classification

Taxonomy is the practice and science of classification. When applied to the field of Integrated Circuits one can get a complete picture of this rapidly evolved field in a span of few minutes.

The most interesting categorization with respect to the theme of this book is Application Specific ICs (known as ASICs) and general purpose ICs. The ASICs generally referred to as Application Specific Products to incorporate the categories like microcontrollers comprises of the following types of ICs:

- System-On-Chip (SoC)
- Semicustom ICs
- Full Custom ICs

Application-specific ICs (ASICs) have the benefit of cramming maximum amount of circuitry for a particular function that leads to space efficiency with cost effectiveness. The main attribute that differentiates the SoCs from others is a mixed blend of many basic functional technologies such as CMOS, bipolar, non-volatile memory, power DMOS, and Micro-Electro-Mechanical System (MEMS) to serve the given application. The full custom IC design is being followed traditionally since the pre-Design Automation era and is still continued with the back up of the state of art Electronic Design Automation (EDA) tools. The main intention of the customers behind choosing the Semicustom paradigm is as follows:

- Design volume is small that poses difficulty in project finance.
- Market lead time forcing the designers to test and validate on the fast prototyping platforms
- Design security is to be maintained to protect the valuable IP

The implementation platforms for the semicustom ICs are the generic chips consisting of reconfigurable arrays of logic. These days even it is a standard practice to go for pre-designed primitives such as RISC processors on these generic chips so as to get the advantage of the completely tested logic blocks around which the system can be built. However, the semicustom approach has an implicit compromise for the specifications as the implementation gets restricted to the predetermined logic array (called as Look Up Tables) laid down by the manufacturers. In the following point in depth account of the specifications called as VLSI design Metrics has been taken.

However the latest trend in the market is System-on-a-Chip (SoC) ASIC designs [22]. This field is highly driven by the EDA tools and with optimized combination of high-speed performance, state-of-the-art packaging technology, and the integration of complex capabilities onto a single chip.

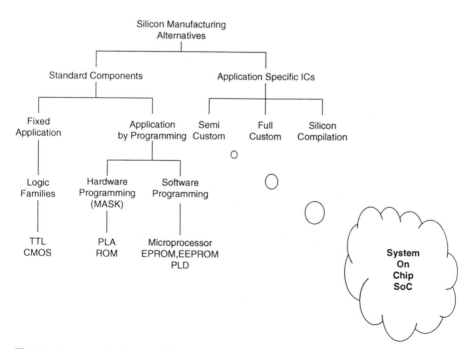

Fig. 1.1 Locating SoC in the VLSI spree

Table 1.2 Comparison of full custom, semicustom and SoC ASIC

Full custom	Semicustom	Soc ASIC
Implementation centered around individual transistors and their interconnects	Implementation centered around the predesigned generic array chips such as FPGA and CPLDs	Designed with the aid of core library, with major focus on RTL design with hardware–software codesign, combination of various technologies and packaging schemes
Spatial and temporal efficiency with the choice of right technology and usage of sophisticated EDA tools for optimization	Required 25–50 times more area leading to spatial inefficiency than other categories	Core based pretested design ensures early analysis of performance with highest level of space and time efficiency
High Non Recurring Cost of Engineering (NRE), costlier for less volume production	Low NRE cost, however more costly for large volume production as large Si area would be wasted for implementation of reconfiguration logic	Cost saving varies widely, depends on production volume and design complexity but definitely less than the semicustom

Way back in the year 1997, the SIA/SRC 1997 Technology Roadmap for Silicon indicated a technical capability to fabricate more than 1 B logic gates, or 64 Gbits of memory per chip by the year 2010. Although most of these predications were achieved much prior to the deadline, a notable prophecy though was not given due importance was "small number of standard chip types". With the growing applications of the FPGAs with their customization for variety of applications manufacturers are forced to curtail the number of standard chips making whatever possible by using the FPGAs. As compared to FPGA based semicustom ASICs, full custom ASIC evidences a longer design cycle and costlier Engineering Change Order. However it has its own undisputed market share due to the attributes of faster performance and lower cost when mass produced. FPGA offers attractive design platform for medium to high volume products. It is seen that many of the successfully tested design in the FPGA environment ultimately gets migrated towards the full custom platform due to power efficiency and minimal critical path.

1.4 System-on-Chip (SoC) Term and Scope

System-on-a-Chip (SoC) is today's most promising revolution happening in the design of integrated circuits due to the possible unprecedented levels of integration. Subsequently, new design methodologies and tools are in demand to address the design complexity and other issues such as simulation, prototyping, verification and test bench generation. The term SoC however is tenuous. Various shades of dimensions of the most evolving SoC are reflected by the definitions given by various sources:

- The System-on-Chip, popularly referred to as SoC, is a chip that holds all of the necessary hardware and electronic circuitry for a complete system. It includes on-chip memory (RAM and ROM), the microprocessor, peripheral interfaces, I/O logic control, data converters, and other components that comprise a complete computer system.
- System-on-Chip technology is the ability to place multiple function "systems" on a single silicon chip, cutting development cycle while increasing product functionality, performance and quality. Even if the semiconductor industry achieved increasing levels of integration, the success of this double integration – processes and functions – involves more capabilities than originally expected [1].
- System-on-a-Chip (SoC) technology is the packaging of all the necessary electronic circuits and parts for a "system" (such as a cell phone or digital camera) on a single integrated circuit (IC), generally known as a microchip. For example, a system-on-a-Chip for a sound-detecting device might include an audio receiver, an analog-to-digital converter (ADC), a microprocessor, necessary memory, and the input/output logic control for a user – all on a single microchip [2].
- The electronics for a complete, working product contained on a single chip. While a microcontroller includes all the hardware components required to process instructions, an SoC includes the computer and all required ancillary electronics. For example, an SoC for a telecom application might contain a microprocessor, digital signal processor (DSP), RAM and ROM. It might also include a graphics processor. The more functions contained within the chip, the more systems can be miniaturized for handheld use with an ancillary reduction in power [3].
- A SoC perception from mixed signal point of views is "System-on-Chip refers to specific conglomeration of chips to be soldered onto a single board. Such Printed Circuit Boards combined on a single platform to work together in a synchronous manner in order to do several processes simultaneously is called System-on-Chip" [14].

Traditionally the System-on-Chip design has been defined as ICs with embedded processors, memory, and other functions that make them much more complex than their basic building-block semiconductor counterparts. However, these days the definition is progressing, to incorporate more application or customer specific products than standard ICs because of the higher levels of integration of the latter as compared to the former. The most appropriate perspective of the SoC looking at the various trends of IP reuse, Hardware-Software codesign and application specific usage is

- A System-on-Chip is an IC designed by stitching together multiple stand-alone VLSI designs (cores) to provide full functionality for an application. Accordingly an SOC comprises of its own processor core (often referred to as an "embedded" processor), and will further comprise one or more cores for performing a range of functions often analogous to those of devices in larger-scale systems [16].

The SoC growth is motivated with the following advantages [19]:

−   Higher productivity levels
−   Lower overall cost
−   Lower overall power
−   Smaller form factor
−   Higher integration levels
−   Rapid development of derivative designs

1.5   Constituents of SoC

The evolution of SoC technology with novel functionality is booming. Shrinking process technologies and increasing design sizes have led to highly complex billion-transistor integrated circuits (ICs) [4, 5]. As a consequence, manufacturers are integrating increasing numbers of components on a chip. A heterogeneous System-on-a-Chip (SoC) might include one or more of the following components:

• General purpose processors cores, digital signal processor cores, or application specific intellectual property (IP) cores
• An analog front end
• On-chip memory (including a selection of ROM, RAM, EEPROM and Flash)
• I/O devices,
• Timing references including oscillators and phase-locked loops.
• Peripherals including counter-timers, real-time timers and power-on reset generators.
• External interfaces including industry standards such as USB, FireWire, Ethernet, USART, SPI.
• Analog interfaces including ADCs and DACs.
• Voltage regulators and power management circuits
• On chip communication protocols

1.5.1   Processor Cores for SoC

Processor core is the heart of the SoC and there exists various implementation strategies such as Control processor with RISC or CISC philosophy, Application Specific High-performance controller, Digital Signal Processor, Audio processor, Network processor and finally the most popular Video processor.

The processor development for the SoC applications is revealing different trends.

There is a great demand for the Customizable processors. These customizable processors are likely to bridge the performance gap between hardwired logic and general-purpose processor. In the current SoC devices, the range of applications of the CPU has been so expanded that the embedded CPU is required to determine the optimum solution for each application. The requirements are getting complicated with the advancements of functions every year [6]. It has been reported that the embedded CPU in a SoC must feature a high processing and transfer capability for operations such as video compression and expansion, by accurately estimating the overall system bus structure and the frequency transfer band of each of its sections. For this purpose, a technique is often used to build the CPU architecture and bus configuration independently and distribute data processing that would impose a heavy load on a single CPU to multiple processors, DSPs and exclusive hardware. As a result of such a solution, the requirements for embedded CPUs are now focused on flexibility and scalability in system construction as well as on processing performance [6]. Even the giants in general purpose processor manufacturing such as Intel has marked its entry into the market for System-on-Chip (SoC) integrated circuits. Known as Atom processor, Intel's low-power processor aims at portable applications, and all set to penetrate mobile and communications markets. Intel is currently working on 15 SoC projects, all tailored for new growth markets, according to Gadi Singer, general manager of Intel's SoC enabling group [7]. Another prevailing trend especially in case of the large System-on-Chip (SoC) design today is to go for multi-processor cores in it. Each of the embedded core will take care of different subsystems such as handling communications, audio or video, alongside one or more host and application processors. However, their debugging is posing more challenges [8].

There are contradictory views regarding the processors for the SoC arena. Designer have gone for the multiprocessor approach with the ASICs and customizable hardware glue logic, to achieve the low power budgeting. However the above mentioned approach suffers from the lack of interoperability and possibility of redundant logic and lacks flexibility.

Researchers also realized that adding a processor with analog functionality helps in reducing the costs, power consumption and gives higher efficiency in performance of the Mixed Signal SoC applications, however at the expense of higher gate count and large memory size for the digital components in the system.

Configurable processor families gain new hardware options and software tool enhancements to appeal to an even wider audience of SoC designers. For instance Tensilica has upgraded its two Xtensa configurable processor families (the Xtensa 7 and Xtensa LX2) with new hardware options and software tool enhancements that make it appeal to an even wider audience of SoC designers Highlights of these capabilities include a new, smaller general purpose register file option, new integer multiplier and divider execution unit options, two new Amba 3.0 bridge options, as well as an easy-to-use new configuration tool that analyses source C/C++ code and automatically suggests VLIW (very long instruction word) instruction extensions that lead to 30–60% improvements in general purpose code performance. These new capabilities provide designers with the most productive configurable processor design environment, with automated features that ensure each processor design is correct by construction.

Table 1.3 Latest SoC processors in market

Sr. No.	Name of the SoC processor/s or platforms	Manufacturer	Applications
1.	Intel® EP80579 Integrated Processor family	Intel	Security, storage, communications, and industrial robotics
2.	MPC8540	Emerson	Embedded computing
3.	TMS320DM335	Texas Instruments	Digital Media System-on-Chip (DMSoC)
4.	e300, e500, e600 and e700	Freescale Semiconductors	Network acceleration, RapidIO interconnect, and Passive Optical Networking
5.	AX110xx family	ASIX	Small form factor, low-cost, embedded and industrial Ethernet applications
6.	EyeQ2	ST Microelectronics	Computationally intensive applications for real-time visual recognition and scene interpretation, and has cabin-grade automotive qualification for use in intelligent vehicle systems
7.	Atlas processors	Centrality Communications	Wireless and GPS-enabled devices such as smart phones, PDAs, and automotive navigation/telematics systems
8.	PVG350MDK	Provigent	Flexible development platform for broadband wireless transmission systems
9.	SMP8654 media processors	Sigma Designs	Advanced Communications' IPv6 set top boxes (STBs)
10.	HiDTVTM Video Processor Family	Trident	High-definition MPEG2 decoding, system processing and video processing features to deliver exceptional video fidelity and system functionality
11.	Tegra 600 and 650	Nvidia	High-definition video on handheld personal computers
12.	LX-001 uP	Grid Connect	Embedded communication or co-processor applications
13.	PMC253	Kontron	32-bit Fieldbus controller for PROFIBUS

Table 1.3   (continued)

Sr. No.	Name of the SoC processor/s or platforms	Manufacturer	Applications
14.	WinPath2 family of processors	WiMAX	WiMAX basestation (BTS) applications
15.	AC494 Voice over packet family of SoC	Audio Code	Echo controller, Adaptive jitter buffer, caller ID
16.	MB93461	Fujitsu Microelectronics Asia Pvt Ltd	Video/audio data compression techniques in the Digital A/V and Multimedia arena
17.	PC3xx family	picoChip	Baseband processors specifically targeted at the fast-growing femtocell market
18.	Geode SC3200	National Semiconductor	Set-top boxes and network terminals
19.	SAF3550 processor	Philips	Cost-effective automotive HD Radio solution
20.	CX9543X	Conexant	Fax, answering machine, speakerphone, and intercom

1.5.2   On Chip Memory in SoC

The main advantage of the on-chip memory is minimal latency. In contrast to main memory, whose latency is hidden by caching, scratch pad memory, a type of on-chip memory currently used in several SoCs, can always be accessed at low latency. In contrast to caches this memory is not managed by the hardware, it is a part of the physical address space. However the usage of the above mentioned on-chip memory has to be managed by the operating system and applications [15]. Another major reason for the growing trend of embedded nonvolatile memory is due to the power saving achieved.

It has been reported [17] that the organization of on-chip memory in embedded processors varies widely from one SoC to another, depending on the application and market segment for which the SoC is deployed. Designer has wide choices ranging from on-chip SRAM, flash, DRAM and even it is possible to judiciously combine the onchip and cache memories to get the optimum performance.

In the recent SoCs, one-time programmable antifuse (OTP) memory macro is becoming common due to the following [48]:

- Most of the SoCs do not require hundreds or thousands of rewrite cycles.
- The altering parameters such as Code storage, calibration tables, and setup parameters can be changed using the memory management algorithms.

- Antifuse OTP has small footprint and hence leads to minimal area-related cost of the die.
- Due to one time programmability, the reliability is good.
- Due to small footprint and hence the lower value of capacitance gives the advantage of reduces precharge and switching power consumption.

The current-generation low-power SDRAM (LPDDR1), available since 2003, still uses relatively high-voltage (1.8-V) I/Os and is limited to low speed (200 MHz) [49]. However, they are soon hopefully going to be replaced by the Low-power DDR2 (LPDDR2), a next-generation low-power memory technology for mobile and embedded designs that offers higher speed, lower- voltage operation, larger capacities and lower pin count than the current generation – and lets nonvolatile memory share the same bus as SDRAM as well.

1.5.3   SoC Buses

In System-on-a-Chip (SoC), the bus architecture plays a vital role to achieve the shorter propagation delay. The conventional buses such as USB, FireWire, Ethernet, USART, SPI are still being used, but suffers severely in the multiprocessor environment owing to their global arbitration mechanism. This further leads to the problems such as congestion, leading to lowered bandwidth and higher latencies not acceptable in systems where deterministic timing is crucial [18].

The AMBA architecture from ARM has been widely adopted in the SoC paradigm due to its various verification attributes such as hierarchy of physical interconnect, unambiguous protocols and growing body of testbench and formal verification infrastructure [20]. Another standard BUS architecture which is not copyrighted is Wishbone. This public domain bus interconnect platform is known for its flexibility for use with semiconductor IP cores. It also exhibits design reuse by alleviating System-on-Chip integration problems by creating a common interface between IP cores [21]. The above mentioned properties of Wishbone improves the portability and reliability of the system, and results in faster time-to-market for the end user.

Researchers have recently introduced a SoC design environment with FPGA based emulation system with automated bus architecture generation. This frame work uses AMBA monitor for debug ability, that samples all bus activities in pin and cycle accurate way. The prototype system can run with clock speed which is close to that of real system and thus assures exhaustive verification and design space exploration. By using the automatic bus generation environment, bus architecture can be generated from the bus specification in an easy and quick way with configurable bus components library. Using this, complicate SOC design processes from the specification to the prototyping system can be done in very small amount of time and effort [22].

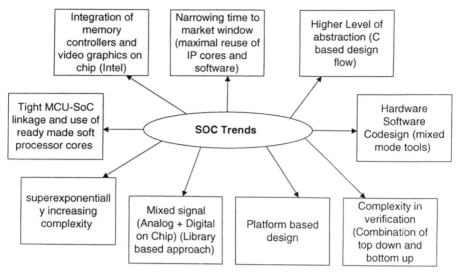

Fig. 1.2 SoC trends

1.5.4 Timing References for SoC

As the modern SoCs migrates towards a much larger scale of integration, it usually needs half a dozen or more clock signals, running at various frequencies. Today's overall timing design usually involves accurate generation, synchronization and distribution of multiple frequencies to drive the operation of different modules such as the core processor, analog front end, i.e., audio codecs, as well as various peripherals. These numerous clocks running at much higher speeds make the noise and interference a big concern for reliable system operation [51].

Timing within all digital and most mixed analog/digital systems comes down to one device –the clock generator. But with mounting system complexity, a single clock oscillator delivering just one digital signal is no longer adequate. These days, systems may require a half-dozen or more clock signals, each at a different frequency [45]. The modern SoCs with Megahertz clock, suffers from large jitter due to variable transmission latency. Phase-locked loops (PLL) derive accurate sample clocks by jitter filtering. The digital implementation of phase detector, loop filter, and clock dividers is straightforward, whereas a digital substitute for analog VCOs is a challenging problem [46]. Therefore the former is preferred in most of the implementations.

More serious issues of concern in modern SoCs with multiple clock signals are noise interference, the mutual interference among clocks and other signals. This could be reduced through good layout and shielding. Different from these methods that passively bypass EMI, changing the clock signal itself through spread-spectrum clocking (SSC) helps actively reduce the EMI generation and is much cheaper and flexible with today's IC capabilities [52].

1.5.5   Voltage Regulator and Power Management Circuits

As the SoCs are progressing towards the low power theme, the voltage regulator and power management subsystem designs are becoming increasingly critical in the applications such as mobile phones due to the GSM noise, occurring with the RF transmitter switching and subsequent draining of high current from the regulator. The remedies suggested are [5]:

- Careful analysis of the PSRR specification over the complete frequency range to compute the sensitivity of the regulator to any possible noise source.
- Selection of a linear regulator with very low drop-out.
- Selecting the regulator to maintain regulation over the full range of the battery voltage.
- Computing the metrics 'effective dropout' rather than the steady state dropout to evaluate the residual noise at the linear regulator output.

1.5.6   On Chip ADCs

ADCs have always played a key role in several application fields of the communications industry. In recent years, the importance of ADCs has been growing mainly due to the developments of the UWB system and Software Defined Radios (SDRs). Nevertheless, in the last 20 years the panorama has consistently changed. As a matter of fact, the requirements for resolution improvements has slowed down, while the demand for higher sampling speeds keeps increasing. Finally, power reduction has become the major challenge in ADCs design [53, 54]. Flash ADC architecture is also known for its high speed operation and is a popular implementation in few SoCs. Here an analog input voltage is simultaneously compared by 2^{n-1} voltage comparators in an n-bit flash ADC. The comparators are, perhaps, the most critical components in a flash ADC. The high-speed comparators are realized with differential amplifiers using bipolar transistors. The comparator realization in the CMOS flash converters consists of auto-zeroed inverter with switched capacitor input.

1.6   Sprawling Growth of SoC Market

SoC is the segment of the semiconductor industry witnessing a very high growth rate and is all set to dominate the market. The main factors fuelling this growth are time-to-market pressures and the ever increasing competition to spur off the

Moore's law with the growing need to embedded more functions in silicon. Way back in the year 2001, the market analysts predicted robust market growth for SoCs [9], with a growth rate as high as an average of 31% a year. However, with the market driven phenomena such as drastic reduction in the overall cycle time of the system with demanding superior performance levels; products migrating to the deep sub-micron complexities, testability issues and time-to-market pressures has resulted in the actual figures much higher than those predicted. The growth rate of the SoC market has witnessed business exceeding 345 million devices in 2003 and 1 billion units in 2004 [10]. United States represents the world's largest SoC market, worth an estimated US$10.4 billion in 2007, as stated in a recent report published by Global Industry Analysts, Inc. Asia (excluding Japan) and Europe are the second and third largest markets respectively. The three regions collectively account for about 80% of the global System-on-a-Chip market [12].

A robust global market growth of SoC with the analysis is presented in a market report 'System-On-Chip – A Global Strategic Business Report' by Global Industry Analysts [11]. A excerpt of the report gives interesting statistics. United States represents the world's largest SoC market, worth an estimated US$10.4 billion in 2007, as stated in a recent report published by Global Industry Analysts, Inc. Asia (excluding Japan) and Europe are the second and third largest markets respectively. The three regions collectively account for about 80% of the global System-on-a-Chip market. The global and regional markets are expected to grow at CAGRs ranging between 20% and 35% through 2010. Market for SoCs Based on Embedded IP is expected to grow at a CAGR of 29% for the period 006–2010, while the Standard Cell based SoCs market is projected to expand at 21%. On the end-use front, consumer electronic devices offer the highest growth opportunity with a CAGR of 31% between 2006 and 2010, followed by automotive applications at 30%. The main reason attributed for the growth is demand for high-speed and low power consuming chips which has expected to increase the growth rate by more than 150% between the years 2006 and 2010.

According to new report, System-on-a-Chip: Technology, Markets, SOC components are being propelled by the following three factors [13]:

- SOC average selling prices are higher than standalone chips, as is to be expected due to integrated functionality.
- SOC unit sales are definitely cannibalizing the consumption volume of the standalone micro-processor unit (MPU), application-specific integrated circuit (ASIC), field programmable Gate Array (FPGA), and digital signal processor (DSP). Therefore, there is the occurrence of SOC devices simultaneously penetrating into the existing markets of most other standalone chips.
- The growth of SOC components has, in fact, led to the birth and popularity of new end-use devices that were hitherto deemed impossible to make or to market. Examples include ultra-small mobile gadgets of the future, ultra-wideband Internet, and certain automobile gadgetry. In essence, SOC has introduced new end-use markets.

1.7   Choosing the Platform, ASIC vs. FPGAs

There is lot of discussion regarding the futuristic evolving platform for the SoC implementation. On one hand although it is agreed that there is and will be always performance gap between ASICs and FPGAs, but still many designers are going for the FPGA based platform. if utmost performance is required. However, the role of full custom design was seen to be diminishing, especially in large SoC systems. The common thing between the ASIC and FPGA is their functionality fixed by the designer and always dedicated for a particular application. Today, full custom ASICs represent a small percentage of the ASIC market because gate arrays, structured ASICs and standard cells turn circuit designs into working chips much faster and at much less cost. Such chips have greatly improved in speed over the years and provide the necessary performance for many applications. The speed advantage of a full custom ASIC is not as relevant as it was in the past. It is used primarily for devices such as microprocessors that must run as fast as possible and will be produced in huge quantities [94].

In nutshell the SoC realization strategies seems to move in the following directions

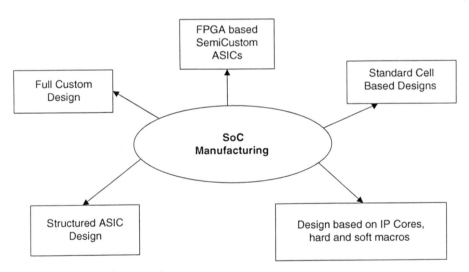

Fig. 1.3 SoC realization strategies

1.7.1   Full Custom Design

Full custom design flow adopts a bottom-up approach in which the design commences with the starts creation of library of cells at lower level of abstraction which then combined to realize the intended system. Generally the primitive cells used for the design have a structures layout interms of transistors along with the layout at mask level supported by the functional model for simulation and floor planning achieved by place and route algorithms. A full custom design empowers the designer with overt control over the physical structure of the design. Here the designer has to carefully work out the logical structure of the design and the rest of the task of converting it into the physical design is undertaken automatically by the EDA tools. Custom design gives much better circuits because it results in structured wiring where key signals have much smaller wire loads [96].

The advantages of 'Full Custom' approach are as follows:

- Smaller die and hence reduced footprint, and lower component cost.
- Possibility of mixed design (analog and digital on the same chip).
- Possibility of high degree of optimization in both the area and performance.
- Integration of pre-designed and pre-verified blocks.

The areas in which an application can benefit from a full custom design are uniqueness of function, performance and cost. The more unique and demanding the application is, the greater the need for a full custom design. On the other hand, almost any design can become less expensive to implement with a full custom design if there is to be a large production volume [95].

The negative points of the full custom design are:

- Time consuming and error prone
- Higher Non Recurring Engineering (NRE) cost
- Support of sophisticated EDA tools required
- Necessitates Experienced Designer with good skills set.

However, the latest work [96], reveals that the full custom SoC can be greatly improved without increasing design time by judiciously employing a number of custom design techniques including floor planning, pre-routing critical signals, tiling data-paths, and generating crafted cells.

1.7.2   Standard Cell Based Design

A standard cell based design approach is gaining popularity amongst today's digital SoC designs. A standard cell is generally a group of transistor along with their interconnect structures, given in a defined layout. The individual cells are treated as the primitives associated with the Boolean logic function or complicated sub systems such as adders, processors etc. These cells are further arranged as a library and made available to the designer as pre-designed pre-verified primitives, which can be readily included in the design as per the requirement. Design with the standard cells makes it possible to globally apply advanced optimization algorithms, which reduce the manual effort required and improve the quality of the synthesized logic during layout. The use of basic standard cell elements reduces complexity to the extent that a complete chip design can be handled flat by layout and test generation tools, removing the need for artificial floor plan boundaries [97]. However, as more analog functions are performed by digital circuits, the sophistication of the standard-cell blocks becomes critical to SoC performance [98]. IC designers are compelled to closely monitor the benchmarks of the cells from the compatibility point of view and fitting them with the other blocks.

1.7.3   Design Based IP-Cores, Hard and Soft Macros

IP cores are nothing but the library of modules (blocks) that can be used in a design. The IP cores can be designed in-house (as is done in this book) or licensed from a vendor. They are the "Macro" structures with specific standard function that can be flexibly being adapted and reused in SoC designs. Although the IPs are ready to use design module their analytical marriage with the existing system poses several issues. One of the important ones is 'Floorplanning'. It plays an vital role in an overall physical design cycle and is instrumental factor for the quality of final layout such as area and performance. The main factors considered in floorplanning are packaging, aspect ratio, routability, timing, and compatibility with the pre-placed macros. The routability and interconnect aspects are very crucial as the wire length has to be estimated during the floorplan phase. There are several methods to estimate the wire length, such as semi-perimeter method, complete graph, and minimum chain, source to sink connection, steiner tree approximation and minimum spanning tree [99]. The real design issue arises when there is requirement of multiple hard, soft and firm cores in the same SoC. In such a circumstances, 'Functional Partitioning' is used to eliminate routing problems when the SoC is laid out on silicon [100].

1.7.4   SoC Through SemiCustom ASICs

The terms "gate array" and "semi-custom" ASIC are interchangeably used by the SoC community and stands for realizing the given design by extensive use of the following platforms:

- Standard Cell Based Platform: Here the design realization is a mix of standard cell libraries, mega-cells, full-custom blocks, system-level macros etc. Although, the design approach is reasonable fast, and economical, it has spatial inefficiency.
- Gate Array Based Design: Here the realization is a mix of channeled (customization of interconnect), channel less (customization of top mask layers) and structures gate arrays (customization of interconnects with well defined structured blocks).
- Programmable SoC Platforms: In this approach, the implementation is based on PLD (PAL, PLA, GAL), CPLDs, FPLDs and FPGAs. With the declining price of

Table 1.4 Comparison of Macros

Comparison feature	Hard macro	Soft macro	Firm macro
Format	Hardware IPs with physical pathways and wiring patterns, mostly in GDSII format	Synthesizable RTL, VHDL, Verilog. System C, Handel C	Netlist (soft macro + layout information in script)
Reconfigurability	Not reconfigurable	More flexible and Reconfigurable	Flexible and portable
Platform	Platform and manufacturing technology dependent, Modeled as a library component.	Platform impendent, incorporable in any manufacturing technology, Synthesizable with several technologies	Platform dependent
Advantages	Optimized blocks interms of power/area/timing.	Editable hierarchical blocks	Optimized for spatial and power Specifications, Flexible and portable
Disadvantages	Non editable IP blocks	Vulnerable to protection, very much dependent on technology and tools, Final specifications depends on the target platform implementation, Timing not guaranteed	Vulnerability same as that of soft macros
Applications	Full Custom ASIC	preferred in SoC implementations	A compromise between soft and hard macros, they are preferred in SoC implementations

the FPGAs from (1 million gates drop from $200+ to <$20) and their predictable development time, accelerated prototyping, programming through simple PC based setup has attracted the SoC community towards them. Since the present book adopts the FPGA based SoC implementation, a detailed analysis regarding the issues pertaining them have been discussed in depth in the following part of the chapter.

1.7.5 SoC Realization Through Structured ASIC

Just for the sake of completeness the structured ASICs are discussed here. They fall somewhere in between an FPGA and a Standard Cell-based ASIC and mainly used to realize mid-volume level designs. Structured ASICs provide a quick-turn

ASIC solution, with lower non-recurring engineering (NRE) cost, and equivalent power and performance to that of an ASIC. They reportedly fill the gap between the 'Cell based Designs' and 'FPGA based Designs'. The products have a much lower cost point and 5X or more density than FPGAs. They can also support lower volumes than a cell-based or full-custom ASIC approach [101]. A Structured ASIC comprises of pre-diffused blocks in standard cell and available metal programmable blocks. Due to certain risks with the structures ASIC platform such as IP-related risks, discontinuing the IP from the vendor, incompatibility of the IPs etc. there was some slowdown in this sector. But they are now thriving due to their positive reward side that offers a substantial portion of the capacity, performance and energy efficiency of cell-based design, but with a lower NRE, a simplified design flow and a faster turnaround time. Thus they now occupy an important space between the simplest cell-based designs and the most powerful FPGA applications [102].

1.8   FPGA Based Programmable SoC

Since the present book emphasizes on the SoC design in the FPGA paradigm, the issues pertaining to it are discussed here in depth.

The FPGA technology is matured and evolved over a period of time so much that it is now preferred platform for the SoCs. With an array of computational elements and the routing wires among them, the capability of reconfiguration on the fly through configuration bit programming has made the designers to use the platform not only for the semi-custom implementation, but also for the full custom SoCs.

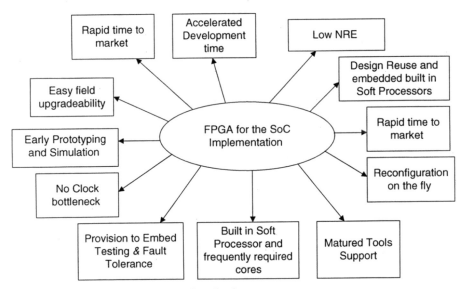

Fig. 1.4   Advantages of FPGA for SoC realization

Apart from the advantages of FPGA for realizing SoC summarized in Fig. 1.4, parallelism can be achieved by intelligent design portioning. Moreover, the prime advantage of designing a SoC to FPGAs for production is the short span from concept to production. Since FPGAs act as their own prototypes, the time needed for each verification iteration is minimized. FPGAs also allow the flexibility of specification changes at anytime during the flow. Of course, serious specification changes could still have a huge impact on schedule but changes can be implemented at any time during the flow without the added cost of silicon re-spins [122].

The rambling growth of the SoC devices based on FPGA has lead to the emergence of platform FPGAs which have rich collection of on chip resources such as CPUs, SRAM, versatile general purpose IO ports, high speed serial links, in addition to the field programmable logic cells. Collection of these functionalities that may be implemented as hard or soft IP cores makes the platform FPGAs extremely flexible reconfigurable SoC devices: they can be customized to a big variety of complex applications by adequately configuring and programming a needed set of available on-chip components. Several programmable logic manufacturers offer platform FPGAs. Altera, Atmel, Xilinx and others propose devices that integrate hardware cores of ARM, MIPS and PowerPC CPUs and/or allow for instantiation of soft processor, DSP and microcontroller cores [123]. For the programs in this book we have opted for devices from the Xilinx Spartan III, in most of the cases and Virtex-II in some cases because their features and evolution path seemed most adequate for our needs when technology choices had to be done.

1.9   Orientation of the Book

1.9.1   Approach Adopted

The comprehension of building System-on-Chip in this book is explored as a tripod approach. To make the tripod of the SoC steady, we have resorted to three important things viz. Handel C based programming methodology, Xilinx FPGAs for realization and prototyping purpose and Soft IP Core of RISC Processor i.e. picoblaze for developing the Embedded applications.

1.9.2   Motivation Behind the Approach

Designing the state of art complex SoCs using the traditional RTL methodology poses several difficulties such as lengthy design and realization cycles because they are at lower level of abstraction Electronic System Level (ESL) tools attempt to address this design complexity problem. The motivation behind the extensive use of the Handel C based design methodology is to give an equal opportunity to the software dominated application sector interested to inculcate their intelligent algorithms on the SoC for whom the some C based design methodology would be more closer to their skill set. Even for the system designers the Handel C based design methodology paves various benefits such as small code size, faster Modeling, Simulation and Verification and exploring the design space effectively apart from the increasing the reusability of the design. As evident from the system design case studies in this book the C-based languages, coupled with block-based approaches are the most efficient approaches (in all sense design efficiency, productivity improvement, abstraction level etc.) and thus these ESL design approaches leads to transaction level models (TLM). In consistent with the reported versions of many designers, Handel-C, based on the ANSI C standard, is a mature, high level language that adds a least of easy to understand hardware-oriented constructs for design implementation and even the concurrency is explicit using the statements such as par (parallel). The important timing issues in SoCs can be handled by rules within the language with the aid of simulation intensive environment using additional types and communications statements such as channel.

The Handel C based design methodology effectively bridges the gap between the behavioral and structural abstractions as shown by means of modified Gajski's Y-chart in Fig. 1.5.

The Handel C cores of the various modules developed in this book are further converted into the EDIF format and then again converted to the bitstream using

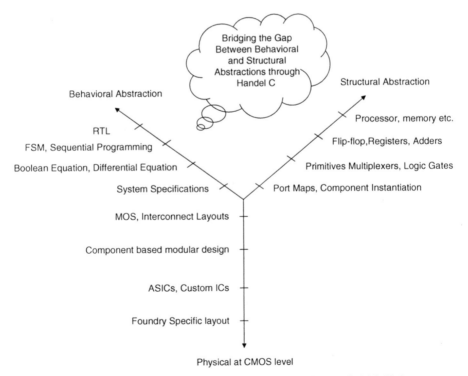

Fig. 1.5 Handel C methodology used in the book justified by using the Gajski's Y-chart

the Xilinx Webpack Version 9.2 for prototyping, realization and verification on the Spartan III FPGA. Other third party tools such as Modelsim have been incorporated in the design flow wherever the need arises.

The most striking know-how given in this book is the interfacing of the Handel C based IP with other soft IP cores using Xilinx EDK tool. This is well illustrated by using an example of 'Adder' interfaced to the Xilinx soft IP processor core 'microblaze'. Looking at the importance of the Network-on-Chip (NoC) paradigm in SoCs, few implementations pertaining to NoCs are presented in the book. The book also demonstrates the accelerated design of few embedded applications using the Xilinx soft processor core i.e. picoblaze.

1.9.3   Setup Used

FPGA kits are available as custom built or commercial and come in a wide price spectrum from different vendors. The designers can test the Handel C cores developed in this book on any custom FPGA prototyping board having standard JTAG connectivity from PC. There are again wide variety of choices for FPGA to PC interfacing. On the PC side the kits can be connected using different options such as RS232, Parallel Port, USB, PCI, Ethernet etc. On the kit side there is standard JTAG interface and a parallel download cable. The board which the authors have used is shown in the Fig. 1.6. However, in order to support the unified hardware and software environment for the soft IP cores developed in Chapter 6, we have used the Spartan 3E starter kit of Xilinx.

Fig. 1.6   Photograph of the setup used for realization of the SoC cores

Chapter 2
Familiarizing with Handel C

Established C-based design flows now provide a rapid, unproblematic design flow from algorithms to FPGA implementation. With the apt coding styles, methodology and design flow, SoC designers can get tangible results from their abstract code. The motivations behind the usage of C based paradigm are ease in SOC design and testing, possibility of working at varying levels of abstraction, effective hardware/software codesign, focused simulation, effective debugging and quick verification. This chapter introduces Handel C a powerful language superset of ANSI-C for designing SoCs.

2.1 EDA Tools i.e. Computer Aids for VLSI Design

Chapter 1 covered the trends in the VLSI technology. It is revealed that with the increasing market demand for ICs with more on-chip functionality at a lower cost and within shortest possible development time span (shorter time to market), the chip developers are forced towards the small feature sizes and this has resulted in exponential increases in the transistor counts. VLSI industry has certain peculiar characteristics, here the manufacturing technology is matured enough to support the increasing transistor count per chip, while the design has seems to be fallen behind resulting in 'design productivity gap'. There are several issues associated with the design productivity gap such as lack of trained HR, availability and usage of the design tools and the right design approach. Since long time, the designers are taking the help of computer based design environment with intelligent tools to accelerate the design speed.

The pre-EDA era ended around 1970, with the design of the first of its kind place and route tool. Prior to this the designer were using computers just for drafting rather than design. In today's industry with phenomenal growth rate and shrinking time to market window the importance of the C, C++ based tools are becoming vital requirements of the VLSI laboratories. C has a very long and rich history of development and evolution. In 1970 Dennis Ritchie from Bell Laboratories created

R.K. Kamat et al., *Unleash the System on Chip using FPGAs and Handel C*,
DOI 10.1007/978-1-4020-9361-6_2, © Springer Science+Business Media B.V. 2009

history with development of 'C' language and using further to write an operating system based on it.

The 'C language has evidenced many evolutions in its lifecycle'. During the 1980s the use of the C language spread widely, and compilers became available on nearly every machine architecture and operating system; in particular it became popular as a programming tool for personal computers, both for manufacturers of commercial software for these machines, and for end-users interested in programming [24]. The 'C' language has grown in different directions with different objectives in mind consistent with the industry requirements. The success of C has been attributed to many factors such as simplicity, portability with small core, ability of generating the space and time efficient codes etc. However, from the hardware transparency point of view the most striking feature of the C which has helped in extending towards the hardware synthesis or VLSI design is its ability to make the programs abstract, facility to build libraries and use them in future programs which supports the industry's latest theme 'Design and Reuse'. The main obstacles designers encountered while extending the C to hardware synthesis is its inherent sequential nature as against the basic concurrency the basis of hardware.

2.2 Background of Hardware Description Languages

Around the same time, when Dennis Ritchie was engrossed with the development of 'C' and writing OS based on it, a major philosophical theme emerged out is using the computer itself as design aids for Electronics and orienting the software world for design activity. In 1972 Ralph Preiss of IBM defined "design automation" as the art of utilizing digital computers to help generate, check, and record the data and the documents that constitute the design of a digital system. Although the notion of the design automation seen from the above definition is more towards the digital system, the concept of describing the circuit and visualizing its functioning before building through simulation based on the underlying mathematical models gave rise to the development of SPICE mainly devoted for the analog world.

The pioneering work in development of hardware description languages was done at Carnegie Mellon University and University of Kaiserslautern, around 1977. The first generation languages are ISP developed at Carnegie Mellon University, and KARL developed at University of Kaiserslautern. The impact of software languages was so much that the developers of ISP couldn't design it very different from them as was seen from the interface used for describing relations between the inputs and the outputs of the design. Naturally it was more for simulate than synthesis. However, the developers of the KARL could orient it more towards VLSI design by including features supporting floor-planning and structured hardware design. Soon after the ABLED graphic VLSI design editor was launched which was centered on the KARL.

Simultaneously parallel development were on towards the hardware end. MMI introduced a breakthrough architecture of the Programmable Array Logic or PAL in 1978. With this the industry spun off the operation of development of HDLs for the programmable sea of gates aiming to reconfigure them with the output of the HDL. Motivated by the PLD architectures, Data-I/O introduced yet another HDL with a trade name ABEL which eased the eventual translation of FSMs to configurable logic.

However, the second generation of commanding and intelligent HDL commenced with the development of Verilog, introduced by Gateway Design Automation in 1985. Regarded as the mother of the modern HDLs soon a series of versions of Verilog along with others such as VHDL started occupying the market share as well as revenue. The VHDL started its development with defense funds but became the default standard of the VLSI chip developers in 90s.

The growth of the VLSI industry fuelled the development of both VHDL and Verilog which have now acquired the dominant position in the electronics industry surpassing the older ones. However, in these days of software dominance, the value of any embedded design largely depends on the quality of the algorithm defining the system functionality and on how fast that algorithm is executed. Typically these algorithms are modeled and refined in C code rather than in VHDL. This is convenient for programming an SoC, but performance requirements will often call for custom hardware to serve as a coprocessor. C-based design flows to programmable logic overcome that issue [41].

2.3   Expressing Abstraction at Higher Levels

There is a constant quest amongst the designers to express the hardware at higher level of abstraction. This notion is based on an alternative to writing HDL by moving to higher level by using the C language. By raising the level of abstraction, algorithms can be implemented without the need for knowledge of the hardware implementation. Celoxica's DK design suite offers a full integrated design environment that enables the user to generate either EDIF netlists or synthesizable RTL to be programmed onto an FPGA. An even higher level of abstraction can be obtained by implementing algorithms in Matlab. Xilinx's AccelDSP allows a design flow to move from Matlab algorithm implementation all the way to down to RTL. The high level of abstraction of Matlab allows for very quick implementations and verification of algorithms [25].

Many high-end designs in the communications or video/image processing industries rely on extremely complex algorithms. The first step in a conventional design flow involves modeling and proving the design functions at the algorithmic level of abstraction, using tools such as MATLAB or plain C/C++ modeling. MATLAB works well for validating and proving the initial algorithm, although many design teams also develop C/C++ models to verify that the whole system meets functional and performance specifications. For subsequent discussion, we'll use the term untimed algorithm to represent those algorithms written either in MATLAB or pure ANSI C/C++ [26].

The constructs added to languages such as Handel-C (which is used in the present book) provide the link to hardware. But the level of abstraction in the code is kept very high to maintain the advantages that come from using C-based code. There are several things that can be added to the design methodology to make it easier to generate hardware from C-based code without having to become a gate-level hardware expert [41].

2.4   Where C Stands Amidst the Well Established HDLs?

Recently there are lots of attentions into the use of C programming language (or its extensions) for describing hardware as well as software with an intention to support hardware-software co-design processes with a single language. One of the hardest things we all need to learn when starting out with HDL is that we're not programming – we're building hardware and arrays of gates. Having done a *lot* of C and applications programming before I started on VHDL and Verilog I can tell you it took a while to shake off the programmer in me and become a hardware developer. Applying general-purpose programming tactics to HDL too often makes too many gates and highly inefficient chip and logic layouts.

Following features of the C based design paradigm makes them a popular choice [41] amongst the VLSI designer community:

- Ease of simulation and testing.
- Effective description of mathematical descriptions of systems.
- Effective verification by using C-based code as the reference model at the start of the design flow and then using this same model to check each subsequent refinement as implementation is completed.
- Co-simulation and possibility to incorporate third party design tools in the design flow. Facility to plug and play models from Matlab to C to RTL to even FPGA hardware within the same simulation environment used to originally design the system functionality.
- Accelerated design cycle to produce a working FPGA prototype of the system.

Although there is no doubt a push in the direction of higher level languages for hardware design evident from the good number of version of C oriented towards RTL output, none of them are established as perfect design platform and moreover as sound industry standard. However, it goes without saying that the future will hold even higher levels of abstraction [42].

2.5   Introducing Handel C

For applications wherein heavy programming is required such as network protocol analysis, the limitations of VHDL come in to the forefront. Languages such as VHDL, ABEL etc. are meant only for hardware description, modeling and simulation and they pose several limitations for programming intensive implementation of complex algorithms. There are set of tools which translate the description from one domain to that of FPGA programmability, many time without compromising the efficiency measured interms of consumption of hardware resources and resulting final specifications such as critical path, timings etc.

The features of Handel C are as follows:

- Innovative language for realizing algorithms in hardware.
- Facilitates architectural design space exploration.
- Powerful for hardware/software co-design.
- Basic core of ISO/ANSI-C, with extensions required for hardware development Inherently sequential but also supports parallel development using the constructs such as 'par'.
- Built in structures to realize flexible data widths, parallel processing and communications between parallel elements.
- Follows simple timing model.
- Provision to specify the width of a data variable.

In the earlier version of the language there was no support for the pointers due to their lack of hardware counterpart. But in the recent revision DK4, the support for pointer has been included.

Amongst all these tools the most poplar are those converting 'C' into a 'logic circuit' or RTL. There are a number of C based languages presently popular in design community viz. Cyber-Cviii, Esterel-Cix, Handel-C, SpecC, and SystemC.

2.6   Top Down or Bottom up?

The HDLs adopt the bottom-up design methodology, in which the individual design blocks are designed and then combined at a later stage to complete the system. The main pitfall of this approach is the late simulation and verification of the overall system due to which the refinements and correction of errors becomes a costly affair.

As there is more thrust on the individual blocks rather than the overall architecture, the hardware-software codesign issues mayn't be dealt in satisfactory manner. The time to market window widens due to serial execution of the expensive steps and thus poses a tradeoff of cost Vs lead time.

The entire design thrust is on the explicit concurrency, which necessitates to pay more cost in terms of design area and timing as well while describing the sequential paradigm. Many a times the novice faces the problems of meta-stability due to inherent 'infinite loops' due to 'switch on condition' constructs. Moreover the designers are more familiar with the sequential paradigm and encounter problems in digesting the hard realities of the delta delays and Perry.

Moreover the focus on the 'structural' description in HDL's makes the divide between the hardware and software codesign more wider. In order to inculcate intel-

Table 2.1   C for VLSI Design Compilers Developed so far

Sr. No.	Compiler	Features
1.	Cones	• Input interms of behavioral models written in C • Output interms of gate-level implementations • Output format interms of standard cells and programmable logic arrays or programmable logic devices • Accelerated design cycle, efficient designs
2.	HardwareC[37]	• Maps behavioral level specification of hardware to a register transfer level description • Focus on concurrent processes, message passing, explicit instantiation of procedures, and templates
3.	Transmogrifier C[39]	• A compiler for simple hardware description language. • It takes a program written in a restricted subset of the C programming language, and produces a netlist for a sequential circuit that implements the program.
4.	SystemC	• Open-source extension of C++ for HW/SW modeling. • Intended for system level verification industry. • Uses C++ class libraries and simulation kernel for creating behavioral and RTL designs. • Supports hierarchical decomposition of a system into modules. • Facilitates structural connectivity between modules using ports.
5.	Impulse C	• Specialized in modeling sequential operations • Compatible to System C and facilitates interprocess synchronization and communication. • Each process treated interms of separate FSM.
6.	C2Verilog[40]	• Converts C algorithms into verilog • Improved logic synthesis results, • Greater control over the compilation process from C to HDL code • Good compatibility and interface with other system-level design automation (SLDA), functional verification, and synthesis tools in the design flow.
7.	Mitrion C[38]	• Uses concept of "Soft-core" processor • Works on abstraction layer between C code and FPGA • Makes use of mapping APIs • Produces output interms of VHDL IP core for the selected target FPGA architectures
8.	Handel-C	• Known for cycle-accurate application development • Clock frequency is limited by the slowest operation • Compiler specialized in efficient analysis, optimization, of code • Provision to get the output interms of VHDL, Verilog, SystemC, EDIF
9.	SpecC[38]	• Extension of ANSI-C. • Supports behavioral and structural hierarchical embedded systems designs • Focus on synthesis and verification • Heavily used for system-level design and architectural modeling.
10.	Napa C[38]	• Language/compiler intended for RISC or FPGA hybrid processors • Uses datapath synthesis technique • Output interms of VHDL, structural VHDL and structural Verilog

Table 2.1 (continued)

Sr. No.	Compiler	Features
11.	Catapult C[38]	• Algorithmic synthesis tool for RTL generation • Outputs RTL from pure C++ program input • Output interms of RTL netlists in VHDL, Verilog, and SystemC
12.	DIME-C[38]	• FPGA prototyping tool • Poor cycle accuracy • Yields higher clock speed designs • Compiler supports pipeline or parallel optimization

ligence on chip and make the hardware evolvable and fault tolerant, the designers should focus more on the abstract level rather than the structural. 'C' based design paradigm makes is in a better position to achieve this goal by ensuring the participation of the software professionals by offering them the bottom-up, abstract and sequential environment in which they are working for quite some time.

The industry experts often analyzes the reasons behind productivity gap. The main arsons are:

- The hardcore EEs relies more on the structural description and thus adopts the bottom up approach with late results.
- Simulation occurs late in the design flow leading to more iterations to eliminate the errors, re-spins and refining of the design.
- With the increasing complexity, in blistering space forcing the verification cycle more stringent.
- The consumer products such as 'set of top boxes', 'mobile phones' are based on mixed-signal chips with implementation of complex algorithms tracking phenomenal permutations and combinations of the event that require designers to examine their operation over few hundred to thousands of cycles to ensure reliability.

In order to address the above challenges, there is a growing trend towards a top down design methodology. The design methodology is based on definite architecture of the chip as a block diagram followed by its simulation, synthesis, verification, optimization, etc. The high level simulation entails the requirements for the individual circuit blocks that guides the circuit design meeting the intended specifications. At last the entire chip is laid out and verified against the original requirements.

System architects badly required in the shrinking time to market window:

The fundamental difference between the C based environment and the HDLs is the basic design approach. The former adopts top-down, while the later relies on the bottom up. Although the former is picking up albeit slowly he reason being scarce design community with the requisite skill set which demands the ability to work as system architects.

A system architect should be well versed with the different constructs and their selective usage besides the art of modeling as the approach is more abstract in nature.

2.7 Handel C: A Boon for Software Professionals

The Handel C was originally developed by Oxford Hardware Compilation Group. The initial development was based on the language Occam with a focused goal of increasing participation of software professionals in synthesis of hardware which they haven't yet explored. The latest version of the Handel C IDE is known as DK4 which is right now supported by Celoxica.

Although there are many C to hardware compilers already being used by VLSI designers, Handel C deserves a special attention as it is one of the most matured C based design platform with built in well thought out extensions for efficient C to RTL conversion. Some of the most striking features of Handel C are [27–35].

- One level of abstraction higher than an HDL.
- Clear, familiar syntax with small and readable code.
- Explicit parallelism using 'par' statement.
- Simultaneous assignment that yields well-defined timing, fast external I/O and simplified pipelines.
- Compiler deals with the state machine generation automatically.
- Easy simulation cycle.
- Facilitates Channel communication.
- Automated place and route, and bit stream creation for Xilinx and Altera tools.
- Designed for compiling programs into hardware images of FPGAs or ASICs.
- Small subset of C, extended with a few constructs for configuring the hardware device and to support generation of efficient hardware.
- Programs mapped into hardware at the netlist level or EDIF level.
- Bit manipulation operators.
- Possibility of parallel processing of single statements or whole modules.
- Shorter simulation time as compared to other hardware simulators.
- Synchronous programming language with fundamental notion of time.
- Level of design abstraction is above RTL but below behavioral.
- Each assignment infers a register and one clock cycle.

For designing functionality of similar complexity, the world of software design is simpler than its hardware counterpart. The possibility of using software design methodologies to create hardware solutions may seem almost theoretical, but there are many real world examples that demonstrates that it resolves many issues inherent to modern IC development. It shortens design time by a factor of 3–4 times, the language and design systems are easily adopted and the small efficient code makes radical system-level changes simple. Handel C is based on the main attribute of the software design paradigm, to raise the level of abstraction so as to describe in the briefest possible way the desired function rather than its underlying structural detail overcomes many of the difficulties and inefficiencies in contemporary hardware design [26].

Researchers have found many interesting features of Handel C in comparison with the other languages such as AHDL [27]. Handel C code yields circuit size or

Table 2.2   VHDL Vs Handel C

VHDL	Handel C
Designed exclusively for hardware engineers	High-level programming language with hardware output to ease the job of software professionals
Aims to crate highly optimized sophisticate hardware	Aims at fast prototyping and algorithmic level optimization rather than sophisticated hardware.
Expects the designer to be familiar with low level hardware (structural) and a background of the gate-level effects of every single code sequence and construct is required while developing the code	Low level problems taken care by complier, designer works at higher abstraction level
Designer has to deal with gate-level problems like fan-in and fan-out or choosing the appropriate type of gates or registers to be used	Frees the designer from low level problems and assists him at algorithmic level
Designer feels jittery to try many implementation strategies.	Fast simulation, many alternate constructs encourages designer to try several alternatives

the number of gates of the same order as that of its counterparts. Unfortunately in case of some RISC architectures or FIFO kind of hardware, it can't give as high clock rate as that of the manually optimized AHDL or VHDL. Speed tweaking is possible and a option regarding the same has been kept in Handel C. However it is better to go for faster FPGA rather than trying for this option as it will defeat the original purpose of Handel C i.e., accelerated design speed. Another feature from the point of view of a novice C programmers is the readability and clarity of the core code without much knowledge of the hardware. That means the C programmers can look at the code from the software perspectives and optimize the same with his knowledge in the software domain rather than wasting valuable time in the hardware details.

One problem the software professional might encounter due to the lack of their hardware expertise is writing the I/O interfaces. The I/O interface is the actual piece of the code written to define the pin assignment per logic block. However once the I/O interface is developed then it can be very well reused for number of routine design cores unless a completely new design problem is encountered.

2.8   Handel C vs. ANSI C

The Handel C was developed to facilitate C to RTL conversion. It is worthwhile to review the subtle difference between the above mentioned duos. The main additions of Handel C to the ANSI C are as follows [42]:

- Parallelism
- Timing
- Interfaces
- Clocks
- RAM/ROM
- Shared expression
- Communications
- Handel-C libraries
- Floating Point component library
- Bit manipulation
- Macro functions for hardware block reuse

There is varying level of support for the various Handel C versions for the following:

- Recursion
- Side effects
- Standard libraries
- Malloc()
- Standard floating point
- Pointers

It should be further noted that some of the above mentioned structures needs to be rewritten to inculcate them in Handel C environment.

The Handel C supports multiple main void in the same workspace unlike the only one in one program in ANSI C. Further the main functions may be hybrid in nature having embedded sequential and/or parallel constructs. Bit manipulation is rather efficient in Handel C as one can declare the integer with variable bits.

2.9 Handel C Design Flow

The IDE for Handel C Design is DK4 suite the licensed information of which may be obtained from the Celoxica website http://www.celoxica.com. The summary of the constructs, statements and other miscellaneous features has been explored in depth in the Handel C reference manual available on line at various URLs. The DK design suite is developed in a manner so as to increase the designer productivity by a familiar IDE environment and the peculiar feature is its in built functions typical of a software development environment.

The design flow is presented below:

The first and foremost step is refining the algorithm to be hardcoded from the view point of its hardware realization. The conventional techniques such as FSM, data path control architecture comes to the rescue of the designer to finalize and fine tune the software version of the algorithm. The designer should further divide the design problem and explore the benefits of the proposed implementation by judicious placement of the modules in hardware and software. Unfortunately the soft-

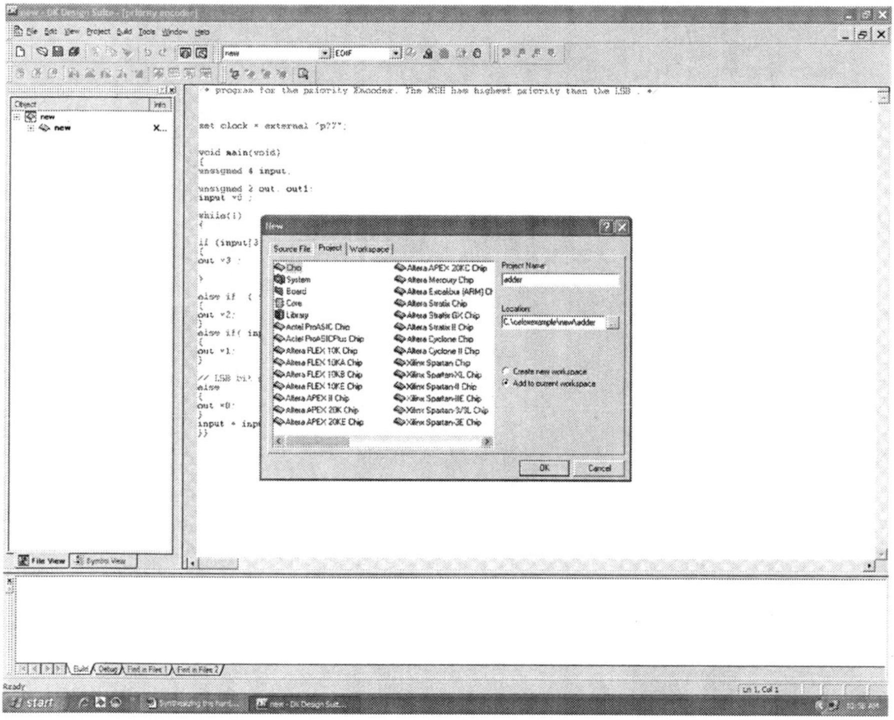

Fig. 2.1 Creation of the project

ware won't help the designer to gain this insight. Then follows the transformation and addition of the constructs for hardware to the C code for the functions to be synthesized to FPGA. At this stage some modifications are required such as changing the floating-point models to fixed-point. The parallel functions can be added at this stage to inculcate concurrency and single clock cycle operations. Some back and forth through these basic design steps will emerge the efficient and optimized hardware-software codesign model of the design problem. The coding in Handel C then can be commenced in the DK4 IDE as given below in the step by step manner:

Step 1: Creating New Project:

A new project pertaining to your design has to be created at the outset. The entire development for this project is done in a workspace in which the source file and header used resides. Creation of a project and the corresponding workspace is shown in Fig. 2.1.

Step 2: Adding or Creating New File:

The source file which could be of text, ANSI C, ANSI C++, ANSI C/C++ header file, Handel C, Handel C header file can be created from the scratch or added if it exits. This is shown in the screenshot in Fig. 2.2. Figure 2.3 Shows the project and source file displayed in the DK IDE.

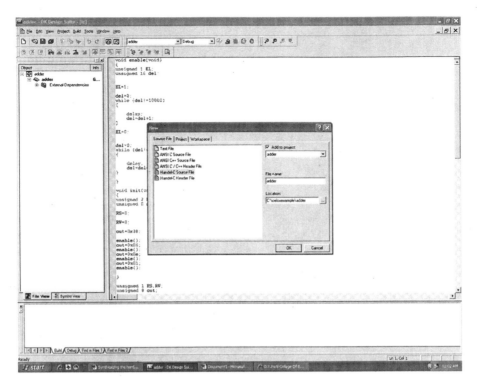

Fig. 2.2   Creating and adding new Handel-C Source file to the project

Step 3: Writing Handel-C Source Code:

The hardware finalized out of the hardware-software codesign process now can be developed in Handel C code as shown in the Fig. 2.4. The software can go independently in C or C++ for development in parallel.

Step 4: Building the Project:

The code developed has to parse through the build step to check the syntax and refinements accordingly to correct the errors. This is shown in Figs. 2.5 and 2.6 respectively.

The hardware primitives utilized are shown in terms of the NAND gates in the watch window. There is further scope for spatial improvements by refining the Handel C code and going back to the hardware-software codesign process.

Step 5: Debugging and Verification:

The effectiveness of Handel C in verification can be realized at this point of design phase. It facilitates easy debussing and verification as shown in Fig. 2.7. The designer can check different variables, clock cycles in the watch windows as shown in Figs. 2.7 and 2.8.

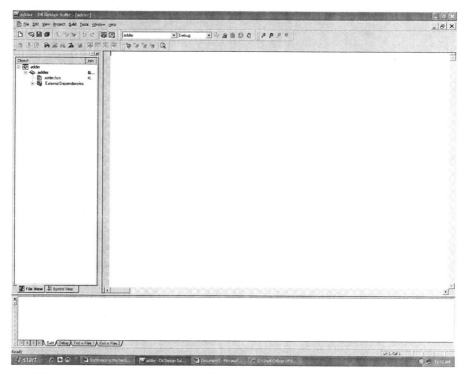

Fig. 2.3   Showing the project and source file displayed in the DK IDE

Step 6: Generating the EDIF:

It involves two substeps viz. selecting the prototyping platform and mapping the
EDIF onto the given platform. The EDIF generated through the Handel C is com-
patible with almost all FPGAs manufactured by the major players in the market. At
this point of time, the designer should be well versed with the FPGA architecture
and the I/O pins to be used. The major task is writing the I/O interface for the chosen
FPGA. Then the EDIF can be created. Once the EDIF is generated, the designer has
top adopt to the third party tools of the respective company for synthesis, place and
route, post synthesis verification and floor plan generation.

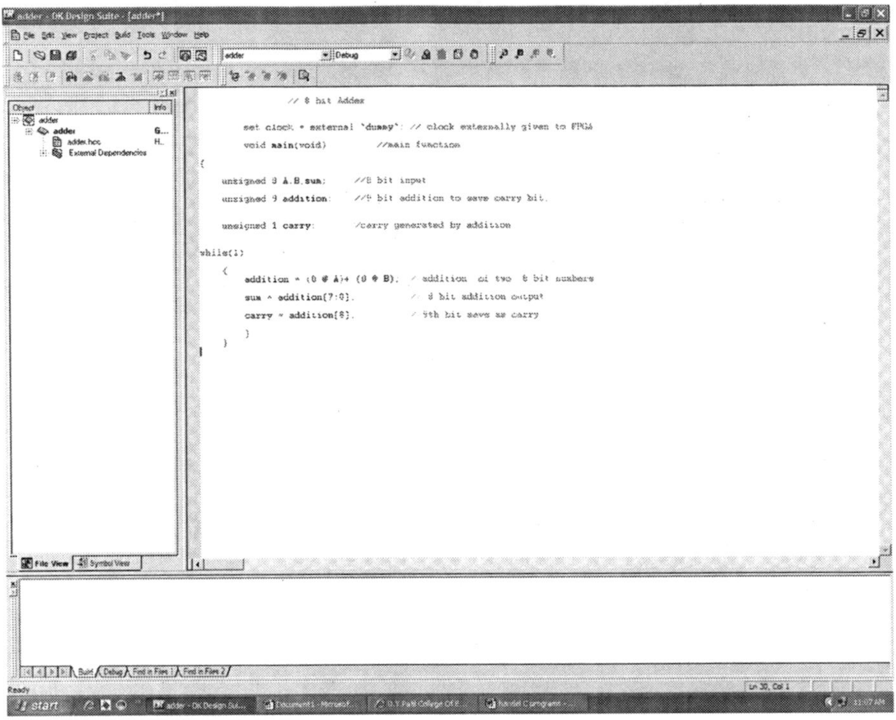

Fig. 2.4 Developing handel C code in DK environment

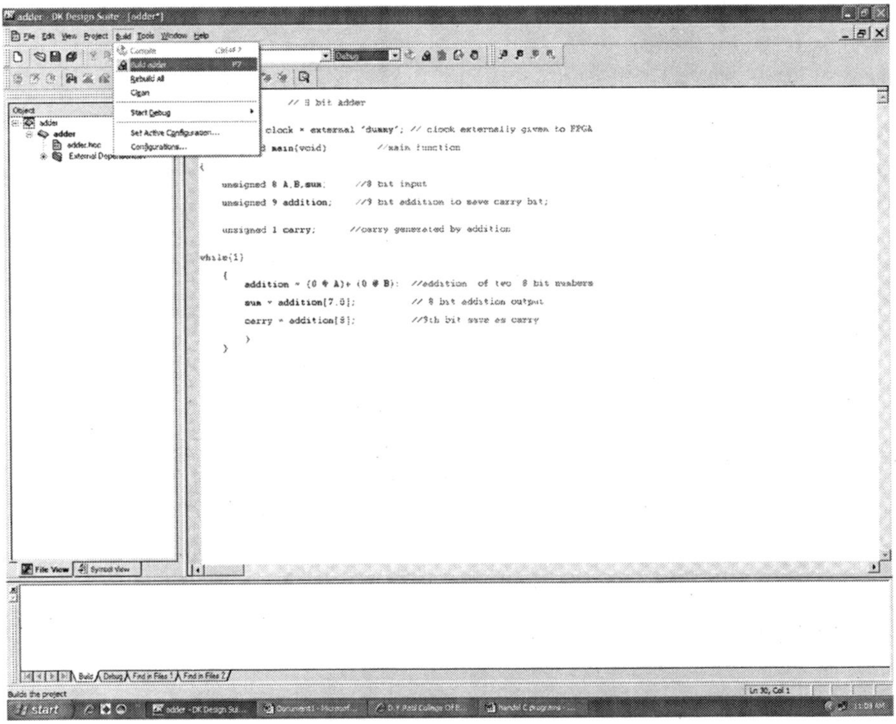

Fig. 2.5 Compiling and building your project

Fig. 2.6 Stimulating the debug process of the project

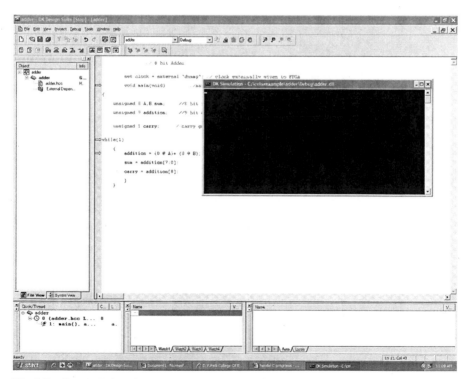

Fig. 2.7 Results being displayed in the simulation window

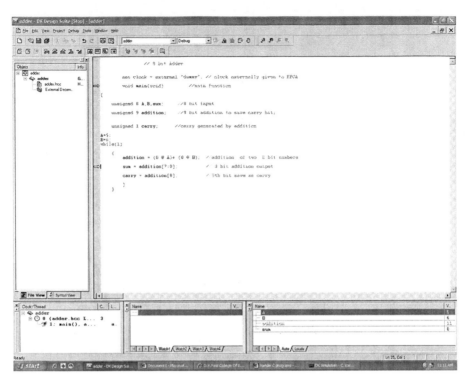

Fig. 2.8   Verification through the watch window

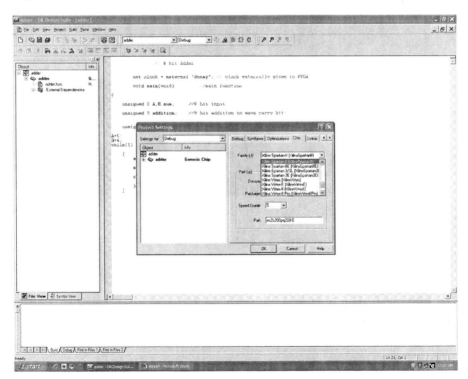

Fig. 2.9   Selecting the prototyping platform by selecting the FPGA family

Fig. 2.10 Generating EDIF for mapping onto the FPGA

Chapter 3
Sequential Logic Design

3.1 Design Philosophy of Sequential Logic

Sequential logic is a type of logic circuit whose output is characterized by not only with the present input but also on the history of the input. The memory aspect of the sequential circuits sometimes poses challenges for the designers. Synchronous circuits are one of the major subclasses of the sequential domain characterized by output transition only with reference to the clock. In general the synchronous sequential circuit is found to be made up of Flip-flops and combinational gates.

However, from the VLSI or ASIC viewpoint the sequential logic circuits are to be dealt carefully as they are the majority implementations in ASICs in different forms such as registers, memory elements, counters, to name a few. The above mentioned logic blocks have been combined in seemingly unlimited combinations to pave interesting applications such as pseudo-random code generators, digital filters, data paths and logic operations. In the present era of cramming more and more functions on the silicon die, the designer needs to take a careful account as regard to the malfunctioning of the sequential logic circuits in the formation stage only.

One of the main reasons behind the malfunctioning of the sequential circuits is designer ignorance towards the "timing windows" around the clock transition owing to the setup and hold times of the circuits. The VLSI aspect regarding the sequential circuits entails the inherent speed bottleneck due to limiting clock frequency imposed by the setup, hold and propagation delay encountered at each individual block.

A typical manual design flow for the sequential circuits mainly comprises of the steps such as drawing the state diagram, transforming these diagrams into the state tables also referred to as the excitation tables per output, deriving the K-map for each output and finally coming out with the circuit diagram of the final version of the circuit. However, this gets simplified with the high level design tools such as Handel C exemplified in this chapter.

R.K. Kamat et al., *Unleash the System on Chip using FPGAs and Handel C,*
DOI 10.1007/978-1-4020-9362-3_3, © Springer Science+Business Media B.V. 2009

Let us start with designing simple logic blocks such as Flip-flops. The Flip-flops have underwent many transitions since their inception in 1919 when the first electronic Flip-flop was invented by William Eccles and F. W. Jordan. Although the functioning interms of logic has remained the same form the beginning when they were popularly referred to as the Eccles–Jordan trigger circuit, consisting of two active elements (radio-tubes), there structural formation has been transformed to suit the power, area and timing metrics of the VLSI industry.

3.2   D Flip-Flop

The D (stands for 'DATA') Flip-flop follows the input, and makes the transitions closely matching to that of D input with a difference of setup, hold and propagation delay times. The D Flip-flop emerged out from the basic requirement of elimination of the race around condition, in the RS Flip-flop by providing only one input to a RS latch. The other input of the RS Flip-flop is derived by inverting the signal to present to the other terminal of the latch. The symbol of the D latch is shown in Fig. 3.1.

Fig. 3.1   D Flip-flop

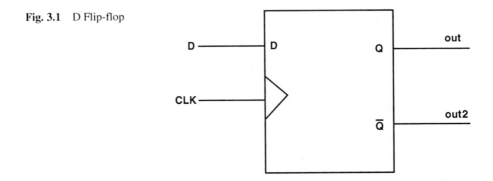

The D FF's are the vital building blocks of many useful applications. The increasing applications of the D-Flip-flop are due to the basic fact that the output depends on only one input as opposed in S-R Flip-flop which has indeterminate state when both inputs are high and JKFF where two inputs decide the output. The meta stability aspects makes the D-Flip-flop as a preferred application in which the designer has to only worry about the loading effects due to rise time, fall time and propagation delay. Literature survey reiterates the same. A design of a D flip-flop hardened to Single Event Upset (SEU) for space radiation environment has been reported by Monnier, T et al. The design hardening technique is based on the use of two D-latch hardened both to static and dynamic SEU by the concepts of high impedance state and nMOS feedback [56]. A basic variation of the D-Flip-flop by modifying it into

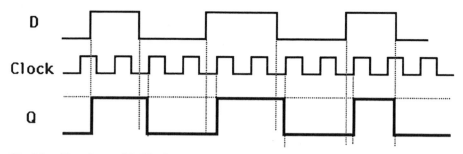

Fig. 3.2 Waveforms of D Flip-flop

a 'Programmable differential D flip-flop' can be used for many useful applications such as temporary storage functions, automatic test equipment, for synchronization of the data signals with timing signals to form a highly accurate timing generator as shown by a patent application [57]. An architecture of asymmetrically faulty D flip/flop under the power line noise, which is based on power flicking reset circuit has been presented and the fail-safe watchdog timer has been reported by Shigeru et al. [58] Even the D Flip-flop can be made to trigger at both the rising and the falling edges of the reference input, thus allowing a lower frequency input to be used while having the advantages of a higher frequency [59].

A program for realizing the D Flip-flop in Handel C is given below:

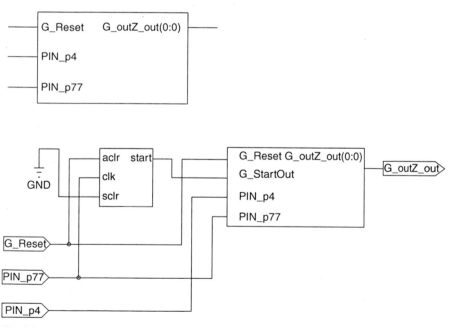

Fig. 3.3 High level realization of the D Flip-Flop on Spartan III FPGA

Program 3.1: Handel C code for D Flip-flop
****

```
set clock = external "p77";
void main ()
{
unsigned 1 D, out, out2;
interface bus_in(unsigned 1 d) inp() with {data = {"p4"}};
//Interface for D input which is given to pin "p4" of FPGA.
interface bus_out() out1(out) with {data = {"p5"}};
//Interface for data output Pin "P5" of FPGA.
while(1)
{
D= inp.d;
Delay;
par
{
out=D;
out2 = ~ D;
}
}
}
```

****

3.3   Latch

The basic difference between the D Flip-Flop and latch is the input tracking mechanism. The former tracks the input only in a synchronous manner while the latter
does the same in an asynchronous manner. In other words the flip-flop is a synchronous version of the latch. By default the latches realized by Handel C are positive
latches i.e. they are transparent with the Clk = 1. Moreover, the latch is level sensitive while a FF is edge sensitive. One more advantage of the Latch over Flip-Flop
is the 'time borrowing', however they are not preferred in the design owing to their
instability as contrasted to flip-flops changing their state only with the clock edges.
Handel C has a built in interface called as **bus_latch_in** that allows the input to be
latched on a condition.

Program 3.2: Handel C code for realizing Latch

****

```
set clock = external "p77";
void main()
{
int 1 get;
int 8 value;
interface bus_latch_in(int 1) inputBus(get) with {data = {"P3"}};
interface bus_out() out1(D) with {data = {"p5"}};
{
get = 0;
get = 1; // Latch the external value
value = inputBus.in; // Read the latched value
D=value; //output the value
}
While(1)
}
```

****

3.4 Realization of JK Flip-Flop

The J-K flip-flop is the most resourceful amongst the basic flip-flops. With the input- tracking character just as that of the clocked D flip-flop it differs interms of the number of inputs, traditionally labeled as J and K. The basic functionality is as shown in the truth table.

Table 3.1 Truth table of JK Flip-flop

J	K	Q	Qbar
1	0	1	0
0	1	0	1
0	0	Previous State	Previous State
1	1	Complement of Q	Complement of Qbar

The origin of the name for the JK flip-flop is detailed by P. L. Lindley, a JPL engineer, in a letter to EDN, an electronics design magazine. The letter is dated June 13, 1968, and was published in the August edition of the newsletter. In the letter, Mr. Lindley explains that he heard the story of the JK flip-flop from Dr. Eldred Nelson, who is responsible for coining the term while working at Hughes Aircraft [60]. With the most predictable behavior under almost every circumstances, the JK flip-flop is preferred for the application with mandatory two inputs. Moreover any flip-flop can be easily constructed from the JK-Flip-flop by proper external wiring. The Handel C program for synthesizing a JK flip-flop is given below:

Program 3.3: Handel C Code for JK Flip-Flop

```
*************************************************************
set clock = external "p77";
void main ()
{
unsigned 1 J,K,Q,Qbar;
interface bus_in(unsigned j) in1() with {data = {"p4"}};
interface bus_in(unsigned k) in2() with {data = {"p5"}};
interface bus_out() out1(Q) with {data = {"p7"}};
interface bus_out() out2(Qbar) with {data = {"p9"}};
while(1)
{
par
{
J=in1.j;
K=in2.k;
}
{
if(J==1 && K==0)
{
Q=1;
}
else if(J==0 && K==1)
{
Q=0;
}
else if (J==0 && K==0)
{
Q=Q;
}
if (J==1 && K==1)
{
Q=~(Q);
}
Qbar=~(Q);
}
}
}

*************************************************************
```

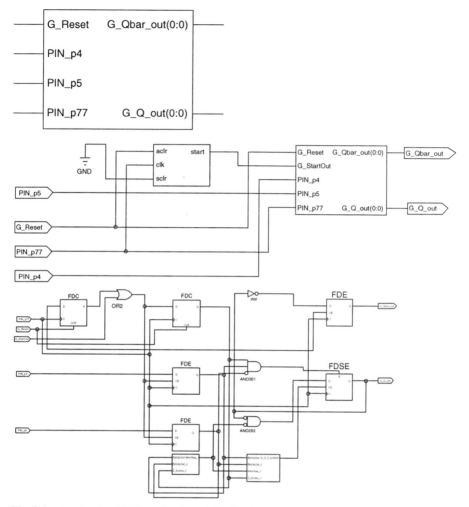

Fig. 3.4   Top level and RTL realization of JK Flip-Flop on Spartan III FPGA

3.5   Cell of Hex Counter for Counter Applications

Counter is a digital electronic device that measures the frequency of an input signal. With little modification in the configuration, it can be used to perform some useful measurements such as period of the input signal, ratio of the frequency of two input signals, time interval between two events and totalizing a specific cluster of events. Many useful application notes regarding the counters may be explored by tracking the references. The know-how of modification in basic counter to work as Normalizing Counter, Preset Counter and Prescaled Counter has also been covered widely [61].

The VLSI viewpoint regarding the implementation of counters is slightly different. Emphasis is given on the replicable hardware, array type design and skew less clock distribution so as to avoid the propagation of erroneous timings. The problems such as meta-stability can be avoided by introducing the buffers in the signal pathways.

(counter count from 0 to f hex)

Fig. 3.5 High level schematic of the Hex Counter

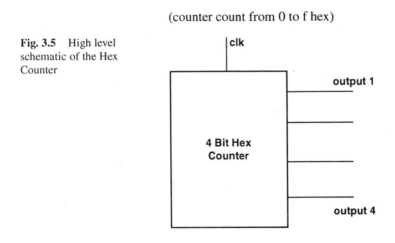

As a sample representative of the counter family, realization of a 'Hex Counter' is given in this section. The handel C program for the Hex Counter is as follows:

Program 3.4: Realizing cell of Hex Counter using Handel C
```
****************************************************************
set clock = external "P77"
        with {extlib = "DKSync.dll",
        extinst = "50",
        extfunc = "DKSyncGetSet"};
                //External clock to FPGA.
void main(void) //main function
{
        unsigned 4 digits;
interface bus_out() output(unsigned 4 out = digits)
        with {extlib = "DKConnect.dll", extinst = "count(4)",
        extfunc = "DKConnectGetSet"};
while(1) //Continues Loop which count from 0 to f
        {
                if (digits == 0b1111)
```

```
                    {
                              digits = 0b0000;
                              }
        else
                    {
                              digits++;
                              }
                }
        }
```

****

The simulation view in Wavweform Analyzer in hex format is as shown in figure.

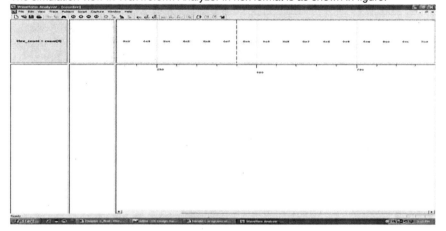

Simulation view in Wavweform Analyzer in stepped format

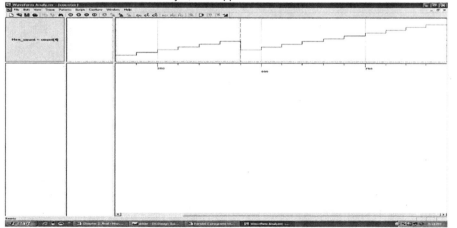

Fig. 3.6 Simulation window of the Waveform Analyzer of DK suite (The *first window* shows the simulation in binary format while the *second one* shows the same in stepped format)

3.6   Realization of Shift Register for SoC

Cascade chaining of the flip-flops generally driven by a common clock with a common preset and clear system leads to a useful data storage and circulation device called as Shift Register. There exists variety of configurations of shift registers as Serial In Serial Out, Serial In Parallel Out, Parallel in Serial Out. The first two types also comes with destructive and non-destructive readouts.

The VLSI strategy for implementation of shift registers is slightly different than that of the discrete electronics. The implementation has to take care of the hold times of the individual flip-flops as the data drifts though them. The spatial inefficiency of the flip-flop rules out their usage for the shift register implementation. Literature survey reveals that the useful implementation is based on the dual port SRAM, with movable read/write pointers to RAM rather than data [62]. The VLSI implementation of the shift registers have largely been used for interesting applications such as tapped delay line, FIFO and LIFO queues in microprocessors and microcontrollers. The dynamic shift registers a popular implementation in 'Design for Testability' adds more value to the implementation by converting the ATPG tests into the multiple, hierarchical scan sequences reducing the test length. Literature survey also reveals development of low power and high speed Bidirectional Shift Register (BSR) architecture with potential applications in hard arithmetic operations, reconfigurable computing and cryptographic implementations [63].

Program 3.5: Realization of universal shift register
```
****************************************************************
/* program for Right Shift */
set clock = external "p77";
void main(void)
{
unsigned 8 s;
unsigned 4 k;
s=0b10000000;
while (1)
{
if (s==1)
{
for (k=0;k<=7;k++)
s=s<<1; //Left Shift
}
else (s==128)
{
for (k=0;k<=7;k++)
s = s>>1; //Right Shift
}
}
}

****************************************************************
```

3.7 LFSR Core for Security Applications in SoC

A Linear Feedback Shift Register (LFSR) is a sequential circuit with the combinational feedback. There are two methods of implementation of LFSR viz. the Fibonacci implementation (consisting of a simple shift register with a binary-weighted modulo-2 sum of the taps fed back to the input) and the Galois implementation (consisting of a shift register, the contents of which are modified at every step by a binary-weighted value of the output stage). The former is also known as simple shift register generator (SSRG), while the latter is popular with other names such as a multiple-return shift register generator (MRSRG) or modular shift register generator (MSRG).

Typical applications of the LFSR implementations are:

- Pattern Generators
- Counters
- Cryptography
- Built-in Self-Test (BIST)
- Encryption
- Bit Error Rate measurements
- Compression
- Wireless communication systems for employing spread spectrum
- Checksums
- Pseudo-Random Bit Sequences (PRBS)

The Handel C code for the LFSR developed in this application resorts to the External LFSR methodology in which all XOR gates are fed sequentially into each other and further end up at the significant bit of the LFSR. It can be easily modified for the Internal LFSR methodology in which the XOR gates are brought inside i.e. they feed into different registers within the LFSR, and thus are not sequential.

Program 3.6 Handel C code for LFSR

```
****************************************************************
set clock = external "p77";
void main ()
{
unsigned 4 out;
unsigned 1 xor_result,input1,input2;
while(1)
{
input1=out[0];
input2=out[1];
xor_result= input1^input2;
out=out>>1;
out=xor_result@out[2:0];
}
}

****************************************************************
```

3.8   Clock Scaling and Delay Generation in SoC

Typically, in frequency divider design, the trade offs are around the maximum operating frequency, power consumption, number of transistors needed and flexibility [64]. PLL and dynamic D flip-flop are the most popular approaches for frequency divider in VLSI arena. One of the possible implementations in Handel C is given below:

Program 3.7: Handel C code for frequency scalar
****

```
set clock = external "dummy";
void main ()
{
unsigned 1 CLKIN, CLKOUT;
unsigned 2 Count;
Count = 0;
CLKIN=1;
while(1)
{
for(Count=0;Count<=3;Count++)
{
CLKOUT =1;
if(Count==3)
break;
}
for(Count=0;Count<=3;Count++)
{
CLKOUT=0;
if(Count==3)
break;
}
}
}
```

****

Build Report:
--------------------Configuration: frequency divide – Debug--------------------
divide.hcc
0 errors, 0 warnings
new
NAND gates after compilation : 318 (15 FFs, 0 memory bits)
0 errors, 0 warnings

3.9   SoC Data Queuing Using FIFO

First In First Out is a general term used to indicate the way of organizing and manipulation of data relative to time and prioritization. It implies 'Queue Processing'
used in the state of art microprocessors, microcontrollers for serving the events in
a timely manner. There are lots of applications of the FIFO technique in buffering,
pipelining, scheduling, networking (in routers, switches) etc. Pioneering work on
the FIFO was done by Peter Alfke way back in 1969 at Fairchild Semiconductors.

The hardware realization of the FIFO logic comprises of set of read and write
pointers, storage and control logic. The popular method of achieving the storage is
by using the dual port SRAM, Flip-flops, latches etc. Literature survey reveals good
number of applications of the FIFO hardware. A design architecture for FIFO Controller, as a module of JPEG2000 Encoder has been reported by Gamad et al. [65].
An implementation of the 'Asynchronous FIFO Synchronizer' for the VLSI Design
and Verification of the Imagine Processor has been reported by Khailany et al. [66]
With the growing application of FIFO, the designers are using the embedded RAM
In FPGAs for realization of FIFO and thus solving the problems of an asynchronous
boundary between clocks [67]. Even implementation of the DMA FIFO helps in
High-Speed Data Acquisition as seen in the application note of National Semiconductors [68]. A design trend is to offer discrete FIFO devices, which cam be selec-
tively glued for the intended application. It is reported that some manufacturers hide
FIFOs within ASICs such as PC chip sets, PCI-to-PCI bridges, network controllers,
and many more. One can also integrate a FIFO into a custom ASIC or FPGA if the
logic gates are available. This approach is especially useful if the designer needs
only a small FIFO and have board-space constraints [69].

An implementation of FIFO in Handel C is given blow so that the designers can
wrap the same wherever required.

Program 3.8: Handel C code for FIFO

```
***********************************************************
set clock = external "P77";
      //Function main
void main(void)
{
      unsigned int 8 data;
static unsigned int 8 LIFO[10] = {23, 46, 69, 92,22,14,18,33,0,9};//for pop
operation
      unsigned index;
unsigned 1 push,pop;
chanin input with {infile = "data.dat"}; // Input from file for push operation
pop=1;
push=0;
      if (push==1&& pop==0)
{
```

```
index=0;
do
{
input ? data; //take input from data file
LIFO[index]= data; //store it on LIFO
        index=index+1; //increment LIFO until 10 location
}
while(index <= 10);
}
else if (push==0&&pop==1)
{
index=0;
do
{
data = LIFO [index];
        index++; //=index+1;
}
while(index >= 10);
}
}
```

****

Build Report:
--------------------Configuration: FIFO – Debug--------------------
fifo.hcc
0 errors, 0 warnings
FIFO
NAND gates after compilation : 1625 (106 FFs, 0 memory bits)
0 errors, 0 warnings

3.10   Implementation of Stack Though LIFO

Last In First Out, a popular implementations in the stack memory refers to the organization of the data elements that can only be added or taken off from only one end, generally in an bottom up manner. The 'Depth First Search' methodology of the LIFO logic has lot of applications in implementation of stack in microprocessors and microcontrollers.

Following Handel C program realizes the LIFO.

Program 3.9: Handel C code for LIFO
```
*************************************************************
/* LIFO (Last in First out); For PUSH operation do not initialize RAM.TAke
input from data. dat file. */
/* static ram unsigned int 8 LIFO[10]; put this as it is For POP operation static
ram unsigned int 8 LIFO[10] = {23, 46, 69, 92,22,14,18,33,0,9}; put this is as
it is. */
set clock = external "P77";
void main(void) // main Function
{
unsigned int 8 data;
static unsigned int 8 LIFO[10] = {23, 46, 69, 92,22,14,18,33,0,9}; //for pop
operation
unsigned index;
unsigned 1 push,pop;
chanin input with {infile = "data.dat"}; // Input from file for push operation
pop=1;
push=0;
if (push==1&& pop==0)
{
index=0;
do
{
input ? data; //take input from data file
LIFO[index]= data; //store it on LIFO
index=index+1; //increment LIFO until 10 location
}
while(index <= 10);
}
else if (push==0&&pop==1)
{
index=10;
do
{
data = LIFO [index];
index--; //=index-1;
}
while (index >= 0);
}
}


*************************************************************
```

Build Report:

--------------------Configuration: LIFO – Debug--------------------

lifo.hcc

0 errors, 0 warnings

LIFO

NAND gates after compilation : 1629 (106 FFs, 0 memory bits)

0 errors, 0 warnings

3.11   Soft IP Core for Hamming Code

Hamming code is a linear error-correcting code (named after its inventor, Richard Hamming) for detection and correction of errors in Digital Communication systems. The Hamming code generation algorithm developed in this application is taken from the web URL: http://www.cs.cornell.edu/Courses/cs414/2007su/slides/ hamming.html. The individual steps are as follows:

1. Marking all the bit positions that are powers of two as parity bits.
2. Remaining all other bit positions are kept for the data to be encoded.
3. Each parity bit calculates the parity for some of the bits in the code word. The position of the parity bit determines the sequence of bits that it alternately checks and skips.
   Position 1: check 1 bit, skip 1 bit, check 1 bit, skip 1 bit, etc. (1,3,5,7,9,11,13,15,....)
   Position 2: check 2 bits, skip 2 bits, check 2 bits, skip 2 bits, etc. (2,3,6,7,10,11,14,15,....)
   Position 4: check 4 bits, skip 4 bits, check 4 bits, skip 4 bits, etc. (4,5,6,7,12,13 ,14,15,20,21,22,23,....) etc. Rest on the similar lines.
4. Setting a parity bit to 1 if the total number of ones in the positions it checks is odd. Setting a parity bit to 0 if the total number of ones in the positions it checks is even

The Handel C implementation of the above steps is given in the following program.

Program 3.10: Handel C code for hamming code IP
****

```
set clock = external "p77";
void main (void)
{
unsigned 5 check1,output1;
unsigned 8 check2,output2;
unsigned 9 check3,output3;
unsigned 10 check4,output4;
unsigned 11 output5;
unsigned 8 input;
unsigned 12 output,check;
unsigned 1 p1,p2,p4,p8;
{
input = 0b10011010;
check1 = 0 @input[3:0];
check2 = input[6:4] @ check1;
check3 = 0@ check2;
check4 = input[7] @ check3;
check = 0 @ check4;
p1 = check[11] ^ check[9] ^ check[7]^check[5]^check[3]^check[1];
p2 = check[10] ^ check[9] ^ check[6]^check[5]^check[2]^check[1];
p4 = check[8] ^ check[7] ^ check[6]^check[5]^check[0];
p8 = check[4] ^ check[3] ^ check[2]^check[1]^check[0];
output1 = p8@ input[3:0];
output2 = input[6:4] @ output1;
output3 = p4 @ output2;
output4 = input[7] @ output3;
output5 = p2@output4;
output = p1@output5;
}
}
```

****

Build Report:
--------------------Configuration: Hamming Code – Debug--------------------
Hamming_code.hcc
0 errors, 0 warnings
Hamming Code
NAND gates after compilation : 888 (130 FFs, 0 memory bits)
0 errors, 0 warnings

Chapter 4
Combinational Logic Design

4.1 Introduction

Combinational logic system is characterized by the basic property that the Outputs being the function only of the current combination of inputs. Thus it exhibits memory less or stateless attributes as well as implies acyclic connection of the primitives. Combinational circuits are known for the ease of debugging as the backtracing through them is uncomplicated owing to the unidirectional nature of the inputs. They are also less prone to glitches, races, hazards and thus more reliable in producing the desired output. Another popular term used for the combinational logic circuits is 'decision making circuits' as they combine the logic functions together to yield a decisive output. A thumb rule for identification of the combinational logic circuit is no level-sensitive or edge-sensitive statement, as well as flip-flops, registers etc.

The VLSI viewpoint regarding the combinational logic circuits throws light on some new techniques. Mapping of the combinational logic functions into the existing primitives of the FPGAs sometimes leads to the inefficiency in terms of the area, delay and power.

4.2 Design Metrics for the Combinational Logic Circuits: SoC Perspective

A general methodology for the FPGA based SoC, combinational logic circuits is technology-specific multilevel logic synthesis, global logic structure synthesis followed by local optimization and finally the technology mapping. The given logic design is expressed in Boolean network form and then the same is subjected to the global logic structure synthesis, and serves as an intermediate description between input specification in HDL or PLA-like format and the final network in a specific target technology [70]. EDA tools based on the 'Multilevel Minimization Algorithms' exist for further reducing the cost of an initially synthesized Boolean network. However, the designs generated by the above mentioned EDA tools after initial synthesis by an algebraic approach are far inferior to manual designs in some cases, owing to the limiting of local node constraints changes rather than global level. More than the importance to the initial structural description of the Boolean Expression, applying certain transformation at the logical level results in minimization of the power consumption. Designers seems to prefer of NAND logic instead of NOR to ensure good performance, whilst reducing power dissipation [71].

With ever shrinking geometries, growing density and increasing clock rate of chips, delay testing is gaining more and more industry attention to maintain test quality for speed-related failures in SoC combinational circuits. The purpose of a delay test is to verify that the circuit operates correctly at a desired clock speed. Although application of stuck-at fault tests can detect some delay defects, it is no longer sufficient to test the circuit for the stuck-at faults alone. Therefore, delay testing is becoming a necessity for today's integrated circuits 0 [72]. SoC designers are also more sensitive to the radiation-induced soft errors which getting worsened in sequential as well as combinational designs on chip. Built-In Soft Error Resilience (BISER) technique for correcting soft errors in latches, flip-flops and combinational logic has now been devised. The BISER technique enables more than an order of magnitude reduction in chip-level error soft rate with minimal area impact, 6–10% chip-level power impact, and 1–5% performance impact (depending on whether combinational logic error correction is implemented or not). In comparison, several classical error-detection techniques introduce 40–100% power, performance and area overheads, and require significant efforts for designing and validating corresponding recovery mechanisms [73].

Finally a word about the hazards and glitches in the combinational logic networks. The basic difference between a glitch and hazard goes on following lines: the former is the momentary change of signals at the outputs while the latter is due to incorrect circuit operation many a times due to the different propagations delays by different paths through which the signal is traversing. The strategy adopted for avoiding both hazards and glitches is to ensure the stability of the signal by using a clock, avoiding asynchronous inputs, and undertaking a thorough analysis of the Boolean expression at the initial level only.

4.3   Core of '2 to 4 Decoder'

A decoder is a decoder is a multiple-input, multiple-output logic circuit that converts coded inputs into coded outputs, where the input and output codes are different. It is a circuit that changes a code into a set of signals. Decoder has good number of applications in code conversion, selective logic activation in devices such as microprocessor buses, memories etc.

From the SoC viewpoint, the design care has to be taken to assign default value for unasserted outputs so as to avoid the excess power consumption. In case of CMOS custom ICs, the decoders can be implemented using the transmission gates and the same results in spatial efficiency. There is also a trend in building the large decoders using the small primitive decoders and decoding with precoded bits.

A Handel C program for the realization of 2:4 decoder is given below:

Program 4.1: Handel C Code for 2 to 4 Decoder IP
****

```
set clock = external "p77";
void main(void)
{
unsigned 1 G,out1,out2,out3,out4;
unsigned 2 In1;
interface bus_in(unsigned 2 a) inp1() with {data = "p3","p4"}};//input pin
interface bus_in(unsigned 1 b) inp2() with {data = {"p5"}};//input pin.
interface bus_out() out(out1) with {data = {"p7"}};//output pin
interface bus_out() outp1(out2) with {data = {"p9"}}; //output pin
interface bus_out() outp2(out3) with {data = {"p10"}};//output pin
interface bus_out() outp3(out4) with {data = {"p11"}}; //output pin.
while(1)
{
In1=inp1.a;
G=inp2.b;
if(In1==0)
{
out1=G;
}
else if (In1==1)
{
out2=G;
}
else if (In1==0)
{
out3 = G;
}
else
{
out4=G;
}
}
}
```

****

Build Report:
----------------Configuration: 2 to 4 Decoder – Debug----------------
Decoder.hcc

0 errors, 0 warnings
2 to 4 Decoder
NAND gates after compilation : 261 (15 FFs, 0 memory bits)
0 errors, 0 warnings

4.4 '3 to 8 Decoder' Using Hierarchical Approach

One of the important aspects of SoC design is hierarchy. The Hierarchical Design is
all about representing the circuit objects encapsulated in a cell definition. Instances
of these cells are then evoked in other cells and thus forms a complex system with
ease of designing. The main advantage of the above mentioned design methodology is ease of debugging and reliable timing as the individual cells are pre-tested.
The following example illustrates design of the 3:8 decoder using two hierarchical
design blocks.

Program 4.2: Handel C Code for 3 to 8 Decoder IP
```
****************************************************************
set clock = external "p77";

macro proc Decoder(In1);
macro proc Decoder1(In1);
void main(void)
{

unsigned 3 In1;

while(1)
{
if(In1<=3)
Decoder(In1);
else
Decoder1(In1);

}
}
macro proc Decoder(In1)
{
unsigned 1 G,out1,out2,out3,out4;
G=0b1;
if(In1==0)
{
out1=G;
```

```
}
else if (In1==1)
{
out2=G;
}
else if (In1==2)
{
out3 = G;
}
else
{
out4=G;
}
}
macro proc Decoder1(In1)
{
unsigned 1 G,out5,out6,out7,out8;
G=0b1;
if(In1==4)
{
out5=G;
}
else if (In1==5)
{
out6=G;
}
else if (In1==6)
{
out7 = G;
}
else
{
out8=G;
}
}
```

****

Program 4.3: Handel C Code for 3 to 8 Decoder IP in non-hierarchical manner using Switch construct

****

```
set clock = external "p77";
void main(void)
{
unsigned 1 G,out0,out1,out2,out3,out4,out5,out6,out7;
```

```
unsigned 3 In1;

while(1)
{
switch (In1)
{
case 0:
{
out0=G;
break;
}
case 1:
{
out1 = G;
break;
}
case 2:
{
out2 = G;
break;
}
case 3:
{
out3 = G;
break;
}
case 4:
{
out4 = G;
break;
}
case 5:
{
out5 = G;
break;
}
case 6:
{
out6 = G;
break;
}
case 7 :
{
out7 = G;
break;
```

```
     }
   }
 }
}
```

****

4.5   Priority Encoder 4 to 2

An encoder in general is a device used to change a group of signals into a code. Priority encoder is a combinational circuit connected by the logic function such that $y=2^x$ inputs and x outputs, the output asserted by the priority of the y inputs. Generally, the most significant bit of the input has the highest priority while the least significant bit has the lowest priority.

Many useful refinements of encoders have been worked out in the SoC devices. Frias et al. have reported a VLSI priority encoder that uses a novel priority look ahead scheme to reduce the delay for the worst case operation of the circuit, while maintaining a very low transistor count. The encoder's topmost input request has the highest priority; this priority descends linearly. Two design approaches for the priority encoder are presented in their paper, one without a priority look ahead scheme and one with a priority look ahead scheme. They have successfully shown that for an N-bit encoder, the circuit with the priority look ahead scheme requires only 1.094 times the number of transistors of the circuit without the priority look ahead scheme. Having a 32-bit encoder as an example, the circuit with the priority look ahead scheme is 2.59 times faster than the circuit without the priority look ahead. The worst case operation delay is 4.4 ns for this look ahead encoder, using a 1-μm scalable CMOS technology. The proposed look ahead scheme can be extended to larger encoders [74].

The Handel C implementation of 4 to 2 priority encoder is given below:

Table 4.1   Truth table of the 4–2 priority encoder

Input				Output	
Input[3]	Input[2]	Input[1]	Input[0]	MSB	LSB
1	X	X	X	1	1
0	1	X	X	1	0
0	0	1	X	0	1
0	0	0	0	0	0

Program 4.4: Handel C Code for 4 to 2 Priority Encoder
****

```
set clock = external "p77";
void main(void)
```

```
{
unsigned 4 input;
unsigned 2 out, out1;
input =0;
while(1)
{
if (input[3] ==1)   //  MSB having higher priority of four bit input.
{
out =3;
}
else if (input[2] ==1)
{
out =2;
}
else if(input[1] ==1)
{
out =1;
}
// LSB bit of 4 bit is lowest priority.
else
{
out =0;
}
input = input +1;
}
}
```

****

Build Report:
--------------------Configuration: Priority Encoder – Debug--------------------
Encoder.hcc
0 errors, 0 warnings
Priority Encoder
NAND gates after compilation : 281 (15 FFs, 0 memory bits)
0 errors, 0 warnings

4.6   Soft IP Core of '7 to 3 Encoder' Implementation

An implementation of 7 to 3 encoder is given below.

Program 4.5: Handel C Code for 7 to 3 Encoder
****

```
set clock = external "p77";
void main(void)
```

```
{
static unsigned 7 inp[1] = {0};
static unsigned 3 out[1] = {0};
if (inp[0]==0b00000001)
{
out[0] = 0;
}
else if (inp[0] == 0b000010)
{
out[0] = 1;
}
else if (inp[0] == 0b0000100)
{
out[0] = 2;
}
else if (inp[0] == 0b0001000)
{
out[0] = 3;
}
else if (inp[0] == 0b0010000)
{
out[0] = 4;
}
else if (inp[0] == 0b0100000)
{
out[0] = 5;
}
else if (inp[0] == 0b1000000)
{
out[0] = 6;
}
else
{
out[0] = 7;
}
}
```

****

Build Report:
----------------Configuration: 7 to 3 Encoder – Debug-----------------
Encoder.hcc
0 errors, 0 warnings
7 to 3 Encoder
NAND gates after compilation : 632 (21 FFs, 0 memory bits)
0 errors, 0 warnings

4.7   IP Core of 'Parity generator' for Communication Applications

A parity bit is generally a redundant binary digit added to the transmitted bit stream for error detection. The error detection mechanism, is based on the checksum of the number of transmitted bits.

There are two variants of parity bits: even parity bit and odd parity bit. 'Even Parity' mechanism ensures checksum of a given set of bits is odd (making the total number of ones, including the parity bit, even). Reverse is true for the odd parity, wherein inclusion of 1 makes the checksum even in case the transmitted sequence is corrupted by noise. A variant of the basic parity code, known as 'Low-Density Parity-Check Code (LDPC code) or Gallager code' is popularly used in digital communication for error detection as well as correction. The other well known applications of the parity codes are:

Ω The SCSI and PCI buses for detection of transmission errors,
• Microprocessor instruction caches for early detection of corrupted data
• Serial data transmission
• Digital recording systems
• Magnetic recording channels
• Modem Links

A parity generator is a classic piece of block code generator circuit that derives the check bits, based on the message bits itself. i.e. Given an n-1 bit data word, it generates an extra (parity) bit that is transmitted with the word based on the 'even' or 'odd' parity predecided mechanism. The Handel C implementation given below is an 8 bit parity generator cell developed using the Ex-OR logic.

Program 4.6: Handel C realization of 8 bit parity generator
```
**************************************************************
// Program for the Parity Generator.
set clock = external "p77";
void main(void)
{
unsigned 8 input;
unsigned 1 out1;
/*interface bus_in(unsigned 8 in1) inp() with {data= {"p3","p4","p5","p7","p9",
"p10","p15","p16"}};
   interface bus_out() out(out1)with {data = {"p17"}};
   */
   while(1)
   {
   input = inp.in1;
   for(input =0; input < 255; input++)
   out1 = (input[0] ^ input[1]) ^ (input[2] ^ input[3]) ^ (input[4] ^ input[5]) ^
(input[6] ^ input[7]);
   }
   }
```

****

Build Report:
--------------------Configuration: Parity Checker – Debug-------------------
parity.hcc
0 errors, 0 warnings
Parity Checker
NAND gates after compilation : 400 (16 FFs, 0 memory bits)
0 errors, 0 warnings

EDIF generating Report:
-----------------------Configuration Parity Checker - EDIF-------------------------
parity.hcc
0 errors, 0 warnings
Parity Checker
NAND gates after compilation : 400 (16 FFs, 0 memory bits)
NAND gates after optimisation : 385 (15 FFs, 0 memory bits)
NAND gates after expansion : 545 (29 FFs, 0 memory bits)
NAND gates after optimisation : 399 (28 FFs, 0 memory bits)
LUTs after mapping : 19 (28 FFs, 0 memory bits)
LUTs after post-optimisation : 20 (25 FFs, 0 memory bits)
0 errors, 0 warnings

4.8   IP Core for Parity Checker and Error Detection for Internet Protocol

The job of 'parity checker' is exactly reverse as that of parity generator. At the receiving end of the transmission, the parity checker makes use of the redundant bits added by the parity generator to detect single-bit errors in the transmitted data word. In order to accomplish this, it has to regenerate the parity bit in exactly the same manner as the generator and after comparing the two parity bits, the disagreement entails the error.

The application developed here is the parity checker for the 'Internet Protocol'. This is a 16 bit parity checker. The basis of the Handel C realization of the parity checker is the mechanism followed in the Internet Protocol. The IP header of the received packet is of 20 bytes size, which is stored in the memory. The byte number 10 and 11 are the parity bits inserted by the parity generator. At the receiving node, these bytes are made zero and the checksum of the remaining header is done. Any disagreement indicates the corruption of the packet in the course of transmission and thus a command is sent for retransmission. Detailed program is given below:

Program 4.7: Handel C realization IP parity Checker
****

```
set clock = external "dummy";
void main(void)
{
```

```
static unsigned 8packet_Header[20];
unsigned i;
unsigned 16 check;
unsigned 16 sum,checksum;
packet_Header[10]=0x00;// Parity Bits
packet_Header[11]=0x00; //Parity Bits
//add 16 bit header
for (i=0;i<19;i=i+4)
{
check =(packet_Header[i] @ packet_Header[i+1]) + (packet_Header[i+2] @
packet_Header[i+3]);
sum = sum +(0@check);
}
while (sum>>16)
sum = (sum & 0xFFFF)+(sum >> 16); // take only 16 bits
checksum = ~sum[15:0]; // complement the result
}
```

****

4.9   BCD TO Seven Segment converter

BCD to Seven Segment Decoder converts a 4-bit binary-coded decimal value, that is the numbers 0–9 coded as 0000–1001, into the code required to drive a seven-segment display. Though 'power hungry' the Seven Segment displays are still one of the popular method adopted for display in PLCs, Microcontrollers and other allied products.

BCD is also a very common and popular for displaying the numeric values in electronic systems. It greatly simplifies the manipulation of numerical data for display by manipulating each digit as a separate single sub-circuit. Designers prefer the BCD system owing to its ease of calculations rather than working with the complicated pure binary system that resorts to conversion and reconversion. Yet another advantage of the BCD system is the smaller code sizes, minimizing the memory and other hardware requirements.

Handel C code developed in this section realizes the BCD to seven segment converter that can be glued on any embedded product. The realization with the Xilinx Spartan III family is shown in the figure. The program is developed for the common anode seven segment displays, with the segment pattern as Dp G F E D C B A.

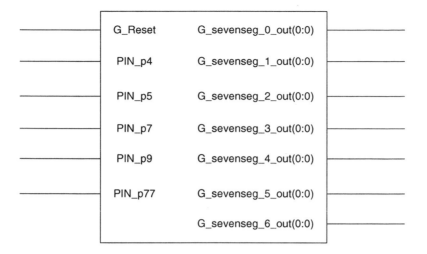

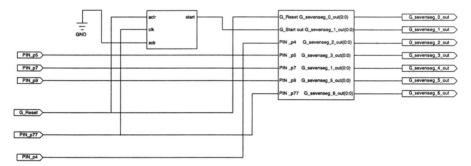

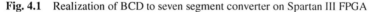

Fig. 4.1   Realization of BCD to seven segment converter on Spartan III FPGA

Program 4.8: Handel C realization BCD to seven segment converter

****

```
// Common Anode
Sequence for Digit is Dp G F E D C B A
set clock = external "p77";
void main(void)
{
unsigned 4 BCD;
unsigned 7 sevenseg;
interface bus_in(unsigned 4 bcd) in1() with {data = {"p4","p5","p7","p9"}};
//4 bit BCD input given to FPGA.
interface       bus_out()      out1(sevenseg)        with       {data      =
{"p16","p17","p18","p21","p22","p23","p29"}};
//seven segment output from fpga.
while(1) // continues loop.
{
BCD=in1.bcd; //input BCD code
```

```
if(BCD==0)
{
sevenseg=0x40;
}
else if(BCD==1)
{
sevenseg=0x79;
}
else if(BCD==2)
{sevenseg=0x24;
}
else if(BCD==3)
{
sevenseg=0x30;
}
else if(BCD==4)
{
sevenseg=0x19;
}
else if(BCD==5)
{
sevenseg=0x12;
}
else if(BCD==6)
{
sevenseg=0x03;
}
else if(BCD==7)
{
sevenseg=0x78;
}
else if(BCD==8)
{
sevenseg=0x00;
}
else if (BCD==9)
{
sevenseg=0x18;
}
}
}
```

****

Build Report:
-------------------Configuration: Seven_Seg – Debug-----Fig2-------
BCD to seven seg.hcc

0 errors, 0 warnings
Seven_Seg
NAND gates after compilation : 1143 (31 FFs, 0 memory bits)
0 errors, 0 warnings

EDIF Generation report:
--------------------Configuration: Seven_Seg - EDIF--------------------
BCD to seven seg.hcc
0 errors, 0 warnings
Seven_Seg
NAND gates after compilation : 1143 (31 FFs, 0 memory bits)
NAND gates after optimisation : 1021 (31 FFs, 0 memory bits)
NAND gates after expansion : 943 (45 FFs, 0 memory bits)
NAND gates after optimisation : 354 (28 FFs, 0 memory bits)
LUTs after mapping : 67 (28 FFs, 0 memory bits)
LUTs after post-optimisation : 68 (26 FFs, 0 memory bits)
0 errors, 0 warnings

4.10   Core of Binary to Gray Converter and Applications

The foremost advantage of the Gray code (named after Frank Gray) over the straight binary code sequence is that only one bit in the code group changes when going from one number to the next one. Thus, the Gray code eliminates the possible error or ambiguity during the transition from one number to the next. Gary codes are also referred to as the reflected binary code, due to their inherent property of differing in only one digit between two successive values. Though originally developed for preventing spurious output from electromechanical switches, today, they are widely used to for error correction in digital communications such as digital terrestrial television, cable TV systems, modem pairs, transceivers etc.

Although algorithms such as 'Shift and Add 3' can be used for the Binary to Gray conversion, we have resorted to the structural EX-OR method which is more temporal efficient. In the algorithmic form it is expressed as:

- Copy the most significant bit.
- For each smaller possible i
- Gray[i] = Binary[i+1] ^ Binary[i]

The Handel C implementation follows:

Program 4.9: Handel C code for binary to gary converter
```
****************************************************************
set clock = external "p77";
void main(void)
```

Fig. 4.2 Binary to gary converter

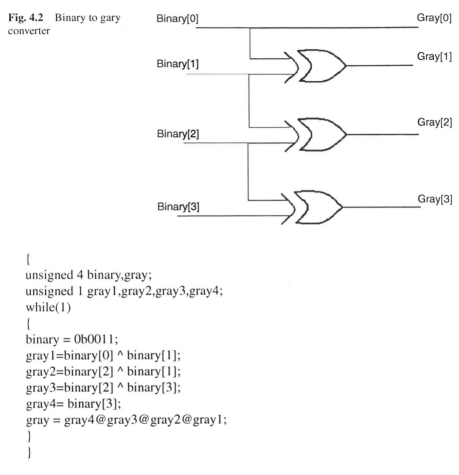

```
{
unsigned 4 binary,gray;
unsigned 1 gray1,gray2,gray3,gray4;
while(1)
{
binary = 0b0011;
gray1=binary[0] ^ binary[1];
gray2=binary[2] ^ binary[1];
gray3=binary[2] ^ binary[3];
gray4= binary[3];
gray = gray4@gray3@gray2@gray1;
}
}
```

****

4.11   Realization of IP core of gray to binary converter

The Handel C realization for the Gray to Binary conversion resorts to the basic algorithm of keeping the Most Significant Bit (MSB) in the Gray code is the same as the binary number and then adding the current Gray code while going from the MSB to LSB (left to right), ignoring the carry. The algorithmic expression can be expressed as:

- Copy the most significant bit.
- For each smaller possible i
- Binary[i] = Binary[i+1] ^ Gray[i]

The same is expressed in the EX-OR logic and written in the Handel C form as shown below:

Program 4.10: Handel C code for realization of IP core of gray to binary converter

```
****************************************************************
set clock = external "p77";
void main(void)
{
unsigned 4 binary,gray;
unsigned 1 binary1,binary2,binary3,binary0;
while(1)
{
binary3=gray[3];
binary2= gray[3]^gray[2];
binary1=binary2^gray[1];
binary0 = binary1^gray[0];
binary=binary3@binary2@binary1@binary0;
}
}


****************************************************************
```

4.12   Designing Barrel Shifters

A barrel shifter is a digital circuit that can shift a data word by a specified number of bits in one clock. cycle. It is an important implementation normally found in SoCs as the basic 'shift' and 'rotate' operations enables to implement multiplication, division, cyclic redundancy checks, cryptography algorithms, floating point arithmetic, bit field extraction, Boolean true/false type structures etc. In most of the SoC implementations the barrel shifter is configured to perform these operations in a stateless, combinatorial manner in a single cycle. It has the ability to rotate and extend signs as well. It accepts 2n data bits and n control signals and produces n output bits.

Program 4.11: Handel C code for barrel shifter
```
****************************************************************
set clock = external "p77";
void main (void)
{
unsigned 4 input,output,rotate;
while(1)
{
```

```
par
{
switch(rotate) //get the data and according the data rotate input.
{
case 0:
{
output=input;
break;
}
case 1:
{
output=input[2]@input[1]@input[0]@input[3];
break;
}
case 2:
{
output=input[1]@input[0]@input[3]@input[2];
break;
}
case 3:
{
output=input[0]@input[3]@input[2]@input[1];
break;
}
case 4:
{
output=input[3]@input[2]@input[1]@input[0];
break;
}
default:
{
output=0b0000;
break;
}
}
}
}
}
```

****

Build Report:
--------------------Configuration: BarrelShift – Debug--------------------
Shifter.hcc

0 errors, 0 warnings
BarrelShift
NAND gates after compilation : 379 (21 FFs, 0 memory bits)
0 errors, 0 warnings

Chapter 5
Arithmetic Core Design and Design Reuse of Soft IP Cores

5.1 Design Reuse Philosophy

As integrated circuit technologies fight to meet and overcome the Moore's law, the quest towards higher performance, greater densities, increasing system complexity and on the top of all, shrinking market window necessities innovation in design methodology. The literature survey reveals that the current, popular design approach, which generally consists of top-down synthesis, software simulation and limited design reuse, is not keeping pace with Moore's Law, which is equivalent to a 59% growth rate. Design productivity has been increasing by roughly 25% per year. This amounts to the growing gap between the designed and manufacturing [75].

On one hand the new EDA tools are trying hard to keep the pace with managing the alarming complexity of the design process and on the other hand the emphasis is given on the IP reuse. The formal definition of the 'Design Reuse' is as follows:

In information technology, design reuse is the inclusion of previously designed components (blocks of logic or data) in software and hardware. The term is more frequently used in hardware development. Design reuse makes it faster and cheaper to design and build a new product, since the reused components will not only be already designed but also tested for reliability. [76]

Design Reuse is a decade old which has emerged with the general usage of embedded processors and standard interfaces mainly to facilitate the tight project schedules and solve the design complexity. It has also helped in reducing the project budget as the SoC industry is hit by the increasing cost of masks.

Some of the prevalent issues in the design reuse are:

- Increasing cost of IP design for reuse due to time to market pressure.
- Intellectual Property Rights of the IP Cores.
- Compatibility of the IP cores with the existing design.
- Lack of documentation regarding the IP.
- Testing and Debugging Challenges as the Testbenches are generally not provided with the IP.
- Optimization of the IP blocks for the intended design as it is being merged as a third party design.

R.K. Kamat et al., *Unleash the System on Chip using FPGAs and Handel C*,
DOI 10.1007/978-1-4020-9361-6_5, © Springer Science+Business Media B.V. 2009

- Timing inefficiency / mismatch when migrating from one technology platform to other.

The guidelines regarding the IP reuse in the SoC industry goes on the following lines:

- Web based interface is being popularized for the IP distribution and reuse.
- This requires creation of the web page for each design and collate the following information
- The RTL source, including the testbench.
- Associated documentation
- Supplementary design notes used during the IP core design.
- Complete listing of the design files in languages like Handel C

Some healthy practices also exist in SoC industry that has fuelled the IP based design reuse as seen from the web based IP repositories below:

Open Cores community and web portal have been founded by Damjan Lampret in October 1999. During 8 years while Damjan Lampret has maintained the web portal, OpenCores became extremely well known in the hi-tech industry. In 2006 over 5000 different companies have downloaded IP from OpenCores. On average 80,000 engineers and others have visited OpenCores web site each month and generated 7.5 million web hits and 2.8 million page views monthly. It is estimated that more than a million engineers have downloaded IP from OpenCores in the first eight years of OpenCores existence (URL: www.opencores.org).

Design and Reuse (D & R) founded by Gabriele Saucier and Philippe Coeurdevey is a web portal for added-value information in the field of electronic virtual component, i.e. IP and SoC. D & R became the worldwide leader as a web and a B2B portal in the IP/SoC field. With its 150,000 page views per month, 15,000 daily updated IP/SOC products descriptions and the on going client/provider matching activity, D & R web stays worldwide unique. 150 companies have signed up a partnership agreement with D & R for providing their latest information to the market through D & R channel and taking the best benefit from its lead service and B2B matching opportunities. D & R has a mission to trigger the IP business (URL: http://www.design-reuse.com/).

Although the reusability of the soft IP cores seems to very simple, it poses several difficulties when it comes to fitting them along with the other modules on the SoC. The first part of the chapter deals with embedding arithmetic cores in Handel C. Later in chapter there is exhaustive coverage of complete realization of the SoC by putting together the soft IP arithmetic core developed in Handel C with the Microblaze, the Soft IP processor core by Xilinx.

5.2   Advantages of on Chip Arithmetic

One of the major problems in executing the arithmetic intensive application on a general purpose computing platform is their speed inefficiency. The interaction between the algorithm and the underlying architecture doesn't match each other which lead to poor throughput. Some times the DSP algorithms have to run in a batch processing mode for an extensive time to get a reasonable solution. There are two possible solutions for bringing the speed efficiency of the arithmetic algorithms the first one is to formulate the algorithm to match the hardware and the second is matching hardware to algorithm. The former approach can be explored by exploiting the equivalence between different operations and then run the modules in parallel. This can be done by extensive analysis techniques such as signal flow or data flow representation. The later approach involves customization of the reconfigurable FPGA hardware to suit the algorithm itself.

This section illustrates the development of the soft arithmetic cores which can be executed in an accelerated manner on the FPGA due to the hardware customization.

5.3   Designing Half Adder in Handel C

A seemingly simple design of the half adder in Handel C is given here with a definite purpose. The Half adder has got many applications. Lutz et al. have reported efficient methods to determine the four usual branch conditions for a sum or difference, before the result of the addition or subtraction is available. The methods lead to the design of an early branch resolver which integrates well with a regular adder/subtracter, adding only a small amount of circuitry and almost no delay. The methods exploit the properties of half-adder form. Sums in half-adder form can be computed very quickly (with the delay of a half adder), yet they have enough structure so that many of the properties of the final sum can be easily detected. The reduced latency for evaluating branch conditions means that an addition or subtraction and a dependent conditional instruction can execute in the same cycle, with a consequent increase in instruction-level parallelism, and improved performance for both single-issue and superscalar processor [124]. There are other reported applications of array of half adders for digital signal processing and multiplier less chips.

The Handel C program of single bit half adder is given below:

Program 5.1: Handel C code for half adder

```
******************************************************************
set clock = external "p77";
void main ()
{
unsigned 1 Input1,Input2,addition,carry;
interface bus_in(unsigned 1 a) inp1() with {data = {"p4"}};//input pin
```

```
interface bus_in(unsigned 1 b) inp2() with {data = {"p5"}};//input pin
interface bus_out() out(addition) with {data = {"p7"}};//output pin
interface bus_out() outp1(carry) with {data = {"p9"}}; //output pin
while(1)
{
Input1=inp1.a;
Input2=inp2.b;
addition=Input1^Input2;
carry=Input1*Input2;
}
}
```

****

Build Report:
halfadder.hcc
0 errors, 0 warnings
adder
NAND gates after compilation : 78 (11 FFs, 0 memory bits)
NAND gates after optimisation : 74 (11 FFs, 0 memory bits)
NAND gates after expansion : 74 (11 FFs, 0 memory bits)
0 errors, 0 warnings

5.4   Full Adder

Like half adder full adder too has been used in various SoC implementations. Burian et al. have reported a new VLSI-suitable hardware implementation of the median filter that uses full adders (FAs) as the basic building block. The proposed hardware structures consist of several stages that exhibit regular and modular structure. It also reduces the hardware requirements and has a faster processing speed, when compared with some other existing hardware implementations [125]. Soudris et al. have also reported systematic graph-based methodology for synthesizing VLSI RNS architectures using full adders as the basic building block. The design methodology derives array architectures starting from the algorithm level and ending up with the bit-level design. Using as target architectural style the regular array processor, the proposed procedure constructs the two-dimensional (2-D) dependence graph of the bit-level algorithm, which is formally described by sets of uniform recurrent equations. The main characteristic of the proposed architectures is that they can operate at very high-throughput rates. The proposed architectures exhibit significantly reduced complexity than ROM-based ones [126].

A Handel C implementation of an 8 bit full adder is given below:

Program 5.2: Handel C code for 8 bit full adder
****
set clock = external "dummy"; // clock declaration

```
void main(void) //start of program
{
unsigned 8 A,B, /*8 bit input
unsigned 9 addition;
unsigned 1 carry;
while(1)
{
addition = (0 @ A)+ (0 @ B);
sum = addition[7:0];
carry = addition[8];
}
}
```

Build Report:
--------------------Configuration: adder – Debug-------------------
add.hcc
0 errors, 0 warnings
adder
NAND gates after compilation : 426 (40 FFs, 0 memory bits)
0 errors, 0 warnings
****

5.5   Ripple Carry Adder

Ripple carry adder is a device for addition of two n-bit binary numbers, formed by connecting n full adders in cascade, with the carry output of each full adder feeding the carry input of the following full adder. The reason to choose for ripple carry adders consists in their power efficiency [127] when compared to the other types of adders. Making an n bit ripple carry adder from 1 bit adders yields a propagation of the CARRY signal through the adder. Because the CARRY ripples through the stages, the SUM of the last bit is performed only when the CARRY of the previous section has been evaluated. Rippling will give extra power overhead and speed reduction but still, the RCA adders are the best in terms of power consumption [128]. With the power optimization, the ripple carry adders are increasingly used in applications involving the arithmetic circuit especially the DSP algorithms in SoC.
   A Handel C program for the ripple carry adder is given below:
Program 5.3: Handel C code for ripple carry adder
****

```
set clock = external "p77";
#define addc(A, B,C) A+B+C
#define add(A, B) A+B
#define carry(A, B) A*B
void main ()
```

```
{
static unsigned 4 Input1,Input2,addition;
interface bus_in(unsigned 4 a) inp1() with {data = {"p4","p5","p6","p7"}};//
input pin
interface bus_in(unsigned 4 b) inp2() with {data = {"p8","p9","p10","p11"}};//
input pin
interface bus_out() out(addition) with {data = {"p12","p13","p14","p15"}};//
output pin
unsigned 1 addi,ca,x,y,z,w;
unsigned 1 a,b,c;
macro expr add= a+b+c;
macro expr carry= a*b;
interface bus_out() outp1(ca) with {data = {"p90"}}; //output pin
while(1)
{
Input1=inp1.a;
Input2=inp2.b;
Input1=14;
Input2=15;
a=Input1[0];
b=Input2[0];
x=add(a,b);
ca=carry(a,b);
addition=0@addi;
a=Input1[1];
b=Input2[1];
y=addc(a,b,c);
ca=carry(a,b);
a=Input1[2];
b=Input2[2];
z=addc(a,b,c);
ca=carry(a,b);
a=Input1[3];
b=Input2[3];
w=addc(a,b,c);
c=carry(a,b);
addition=w@z@y@x;
}
}
```

****

5.6   Booth Algorithm and its Realization on FPGA

The salient features of the Booth algorithm is its encoding scheme to reduce number of stages in multiplication. The algorithm performs two bits of multiplication at once and requires only half the stages; each stage being slightly more complex than simple multiplier. It is the standard technique used in chip design, and provides significant improvements over the 'long multiplication' technique [128].

Handel C program for the Booth algorithm is given below:

Program 5.4: Handel C code for Booth algorithm
****

```
//Booth Algorithm for Multiplication of two number.
set clock = external "p77";
void main (void)
{
unsigned 4 in1,in2;
unsigned 7 out,outf;
unsigned 4 out1;
unsigned 5 out2;
unsigned 6 out3;
unsigned 7 out4;
in1=0b0011;
in2=0b0010;
if (in2[0] ==0)
{
out1 = 0;
} else
{
out1=in1;
}
if(in2[1] == 0)
{
out2=0;
}
else
{
out2=in1@0;
if(in2[2]==0)
{
out3 =0;
}
else
out3=in1@0;
}
if(in2[3]==0)
```

```
{
out4 = 0;
}
else
{
out4 = in1@0;
}
out = (0 @ out1) + (0 @ out2) + (0@out3)+ out4;
outf=out;
}
*************************************************************
```

Build Report:
-------------------- Configuration: Booth Algorithm – Debug --------------------
Booth.hcc
0 errors, 0 warnings
Booth Algorithm
NAND gates after compilation : 982 (59 FFs, 0 memory bits)
0 errors, 0 warnings

5.7   Building ALU in Handel C

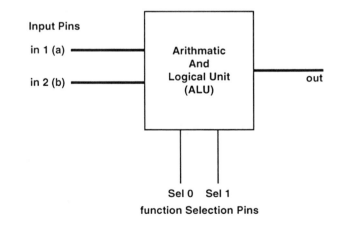

Fig. 5.1 Functional
block diagram of ALU

Sel0	Sel1	Function to be perform by ALU
0	0	Addition (in1 + in2)
0	1	Subtraction (in1 – in2)
1	0	Multiplication (in1 * in2)
1	1	Division (in1 / in2)

Program 5.5: Handel C code for 8 bit ALU

****

```
set clock = external"p77";
unsigned 8 add(unsigned 8 a, unsigned 8 b); //function declaration
unsigned 8 sub(unsigned 8 a, unsigned 8 b); //function declaration
unsigned 8 mul(unsigned 8 a, unsigned 8 b);
unsigned 8 div(unsigned 8 a, unsigned 8 b);
void main (void)
{
unsigned 8 a,b,ad,mu,su,di,out;
unsigned 2 sel;
interface bus_in(unsigned 8 in1) inp1() with {data = {"p3","p4","p5","p6","p7",
"p8","p9","p10"}};//input pin
interface bus_in(unsigned 8 in2) inp2() with {data = {"p11","p12","p13","p14",
"p15","p16","p17","p18"}}; //inputpin.
interface bus_in(unsigned 2 select1) inp3() with {data = {"p19","p20"}};//input
pin
interface bus_out() out1(out) with {data = {"p21","p22","p23","p24","p25",
"p26","p27","p28"}}; //output pin
while(1)
{
a = inp1.in1;
b= inp2.in2;
sel=inp3.select1;
di=div(a,b);
ad= add(a,b);
mu= mul(a,b);
su= sub(a,b);
if (sel==0)
{
out = ad;
}
else if (sel == 1)
{
out = su;
}
else if (sel ==2)
{
```

```
out = mu;
}
else if (sel == 3)
{
out = di;
}
}
}
unsigned 8 add(unsigned 8 d,unsigned 8 e)
{
unsigned 9 temp;
unsigned 8 c;
unsigned 1 carry ;
temp= (0@d)+(0@e);
c=temp[7:0];
carry=temp[8];
return(c);
}
unsigned 8 sub(unsigned 8 s,unsigned 8 p)
{
unsigned 8 c;
c= s-p;
return(c);
}
unsigned 8 mul(unsigned 8 t,unsigned 8 u)
{
unsigned 8 c;
c= t*u;
return(c);
}
unsigned 8 div(unsigned 8 x,unsigned 8 y)
{
unsigned 8 c;
c= x/y;
return(c);
}
```
****

Build   Report:
--------------------Configuration: ALU- Debug--------------------
alu.hcc
0 errors, 0 warnings
book
NAND gates after compilation : 6390 (188 FFs, 0 memory bits)
0 errors, 0 warnings

5.8   Xilinx EDK Interface with Cores Developed Through Handel C

Xilinx EDK (stands for Embedded Development Kit) is a design environment for developing embedded systems using Xilinx FPGAs and it facilitates creation of hardware architecture, compilation of software, integration of software into the hardware by writing driver and finally verification of the RTL. XPS (Xilinx Platform Studio) is the GUI that integrates all EDK processes.

The components of the EDK are as follows:

- Processors (PowerPC, MicroBlaze)
- Interconnect (PLB, OPB, FSL, etc.)
- Memories (BRAM, DDR)
- Peripherals (UART, Ethernet, Custom cores)

As one can see from the list of the components almost any SoC can be realized by using the EDK environment. The missing components can be added through the custom cores.

We are giving here a step by step discourse for building the configurable SoC using the Xilinx EDK as there are lot of queries and apprehensions from the design community and the flow is not yet clearly given in the literature. Through this book

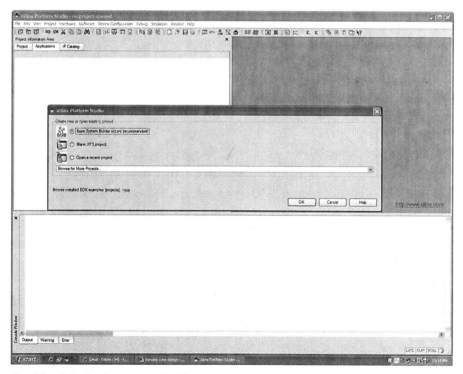

Fig. 5.2   Selecting base system builder

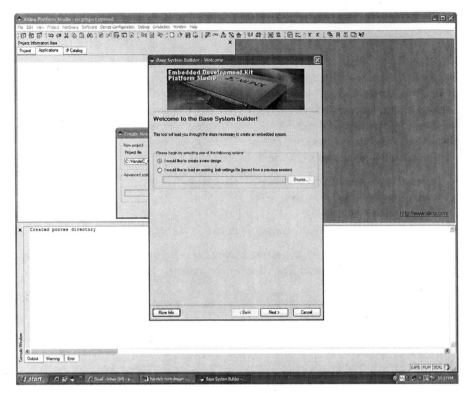

Fig. 5.3   Building project from scratch / on old module

we are empowering the designers to build their SoC by making use of the EDK and
the readily available soft IP cores given here in the form of Handel C code.

5.8.1   Design Problem

In this design a soft IP core of 32 bit adder is developed on Spartan 3E FPGA
which is intended to be interfaced to the microblaze using PLB bus. A data stream
is passed to the adder soft IP core and the result is acquired by the microblaze over
the PLB. Further the result is displayed on PC hyperterminal using the UART interfaced to the microblaze.

The step by step implementation is as follows:

5.8.2   Design Flow

As shown in Fig. 5.1, as soon as the new project is opened, the designer has to select
the base system design which is shown in Fig. 5.2. The EDK then asks the use as

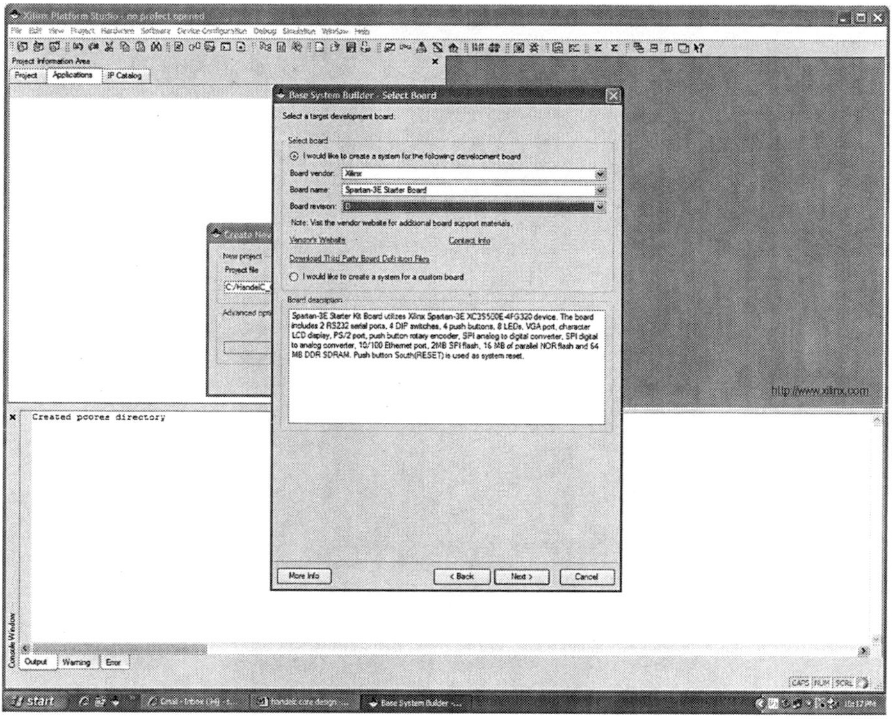

Fig. 5.4 Selecting the Xilinx spartan 3E board

Table 5.1 Files created by the XPS as the SoC is built

Name of the File	Explanation
Base System Builder	Selection of working base system
MHS	Generation of Netlist from Hardware Description
MSS	Software converted into Libraries and drivers
Sim Gen	Generation of Simulation Models
Xilinx Microprocessor Debugger	Debugging information for the microprocessors
BitInit	Generation of bitstream for configuring FPGA

regards to building the project from the scratch or on a previous project. In case the designer wants o build the project on his/her old module then he/she can go for the. BSB file.

As shown in Fig. 5.4, the designer then has to choose the Xilinx FPGA board. The designer can very well go for FPGA boards other than Xilinx, with only additional burden to develop the pin configuration or UCF file for it.

The next step is selection of the available Soft IP processor core. The Spartan 3 FPGA supports only Microblaze by default. Power PC is available for Virtex FPGA.

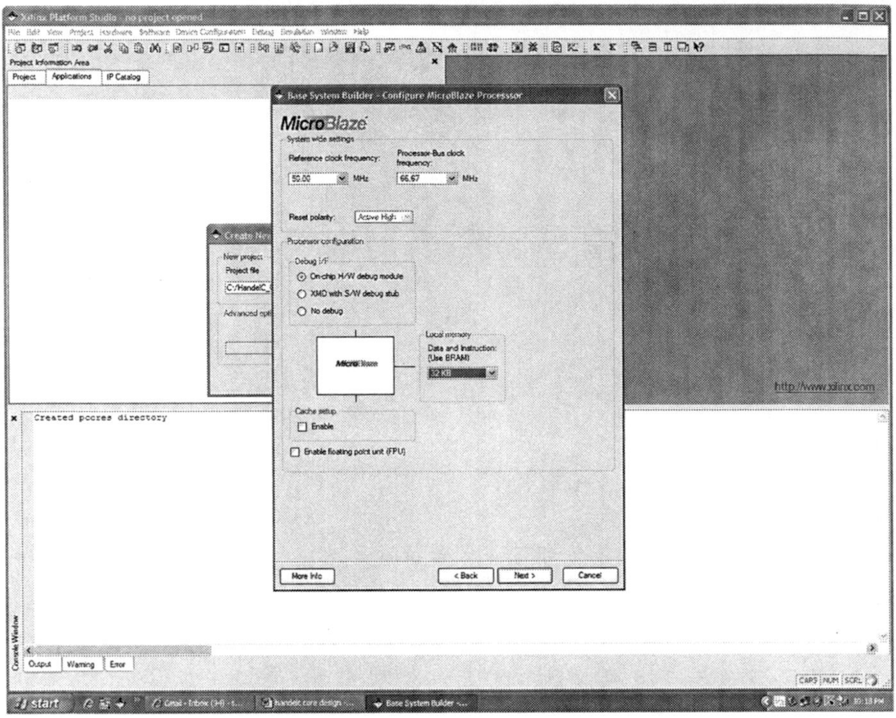

Fig. 5.5 Selecting the soft IP processor

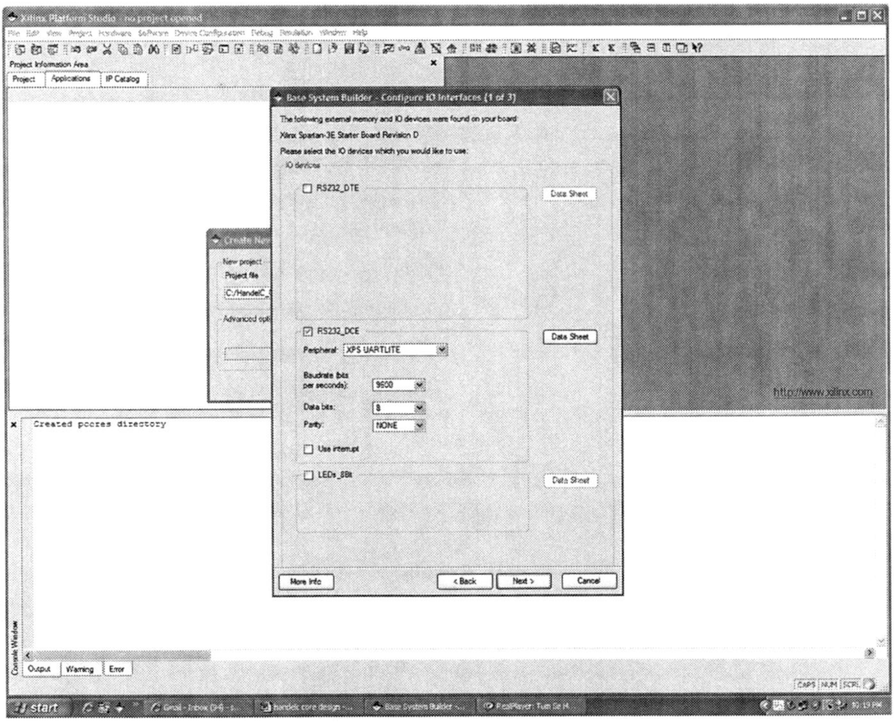

Fig. 5.6 Selecting the UART with appropriate configuration

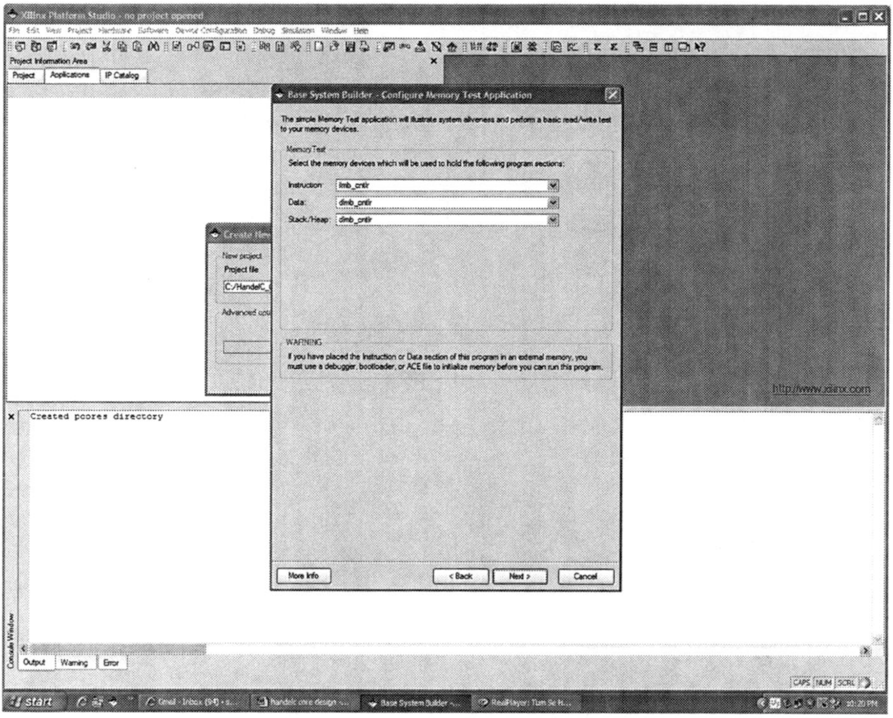

Fig. 5.7 Configuring the UART as standard Input/Output device

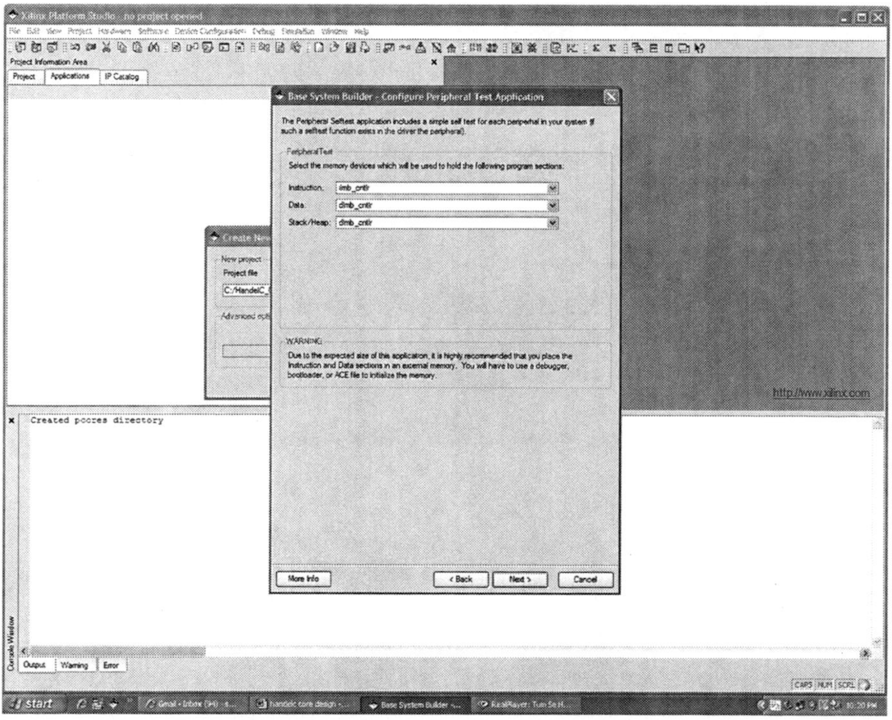

Fig. 5.8 Test bench generation for the peripherals

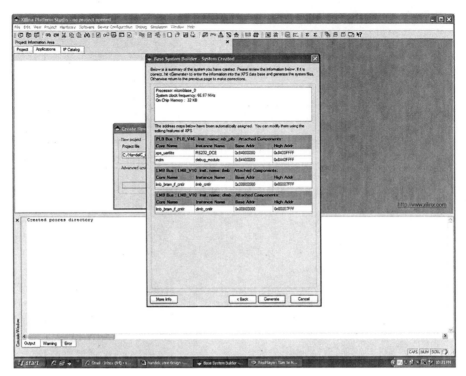

Fig. 5.9 EDK showing the specifications of the system

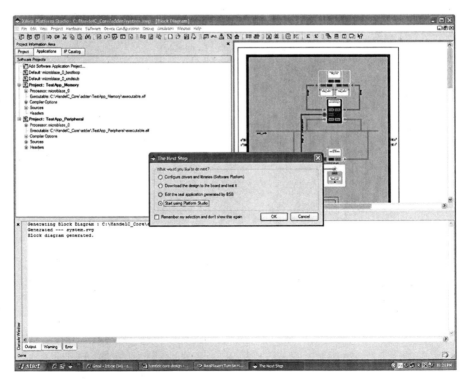

Fig. 5.10 Starting platform studio

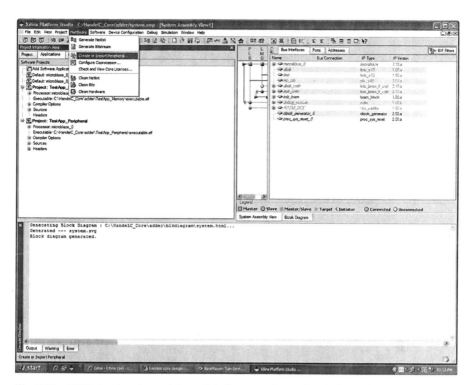

Fig. 5.11 EDK showing the modules and their connections using the buses. The designer can start importing the peripherals as well in this step

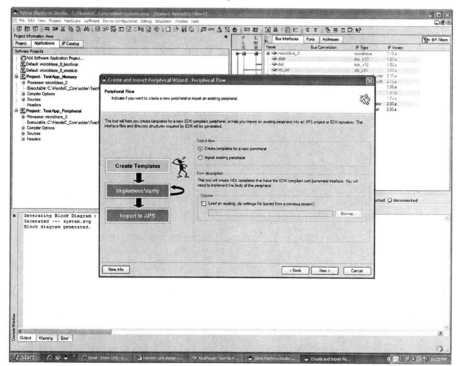

Fig. 5.12 Showing creation and implementation of new peripherals

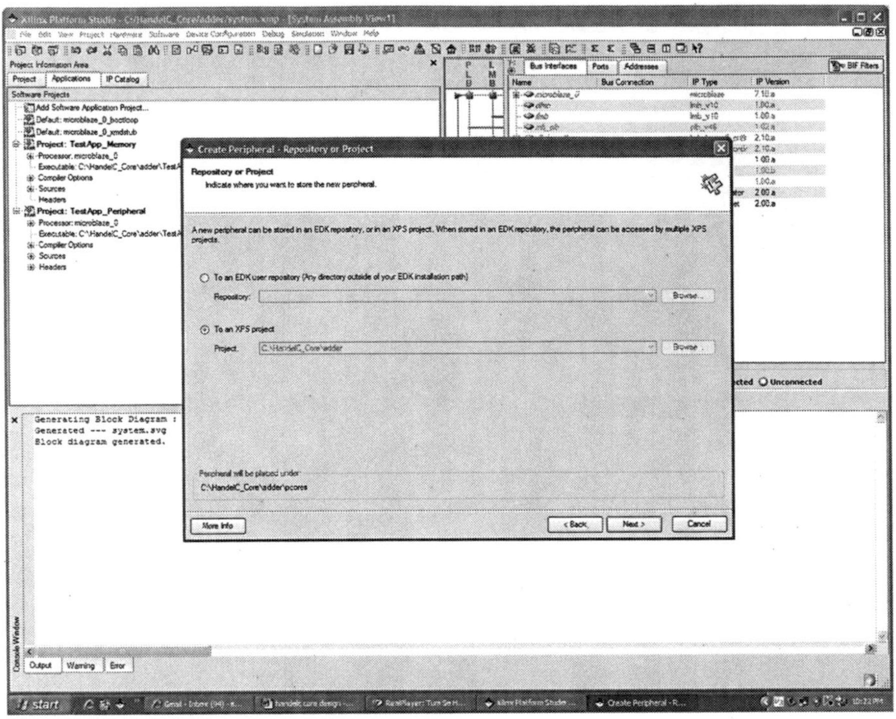

Fig. 5.13 Linking the peripheral to the project by specifying path

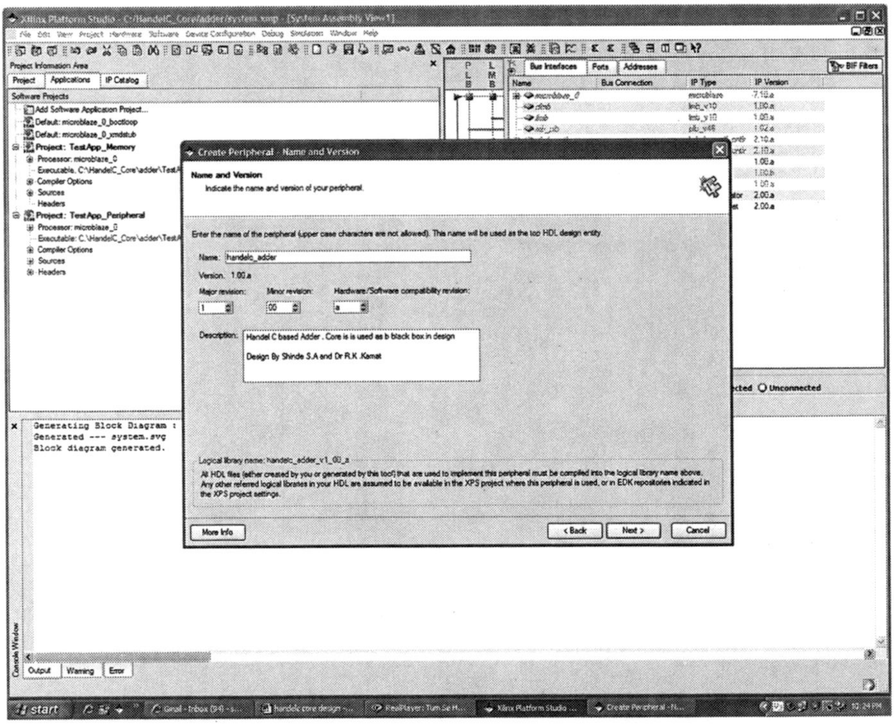

Fig. 5.14 Specifying the Core Name and other information such as revision etc

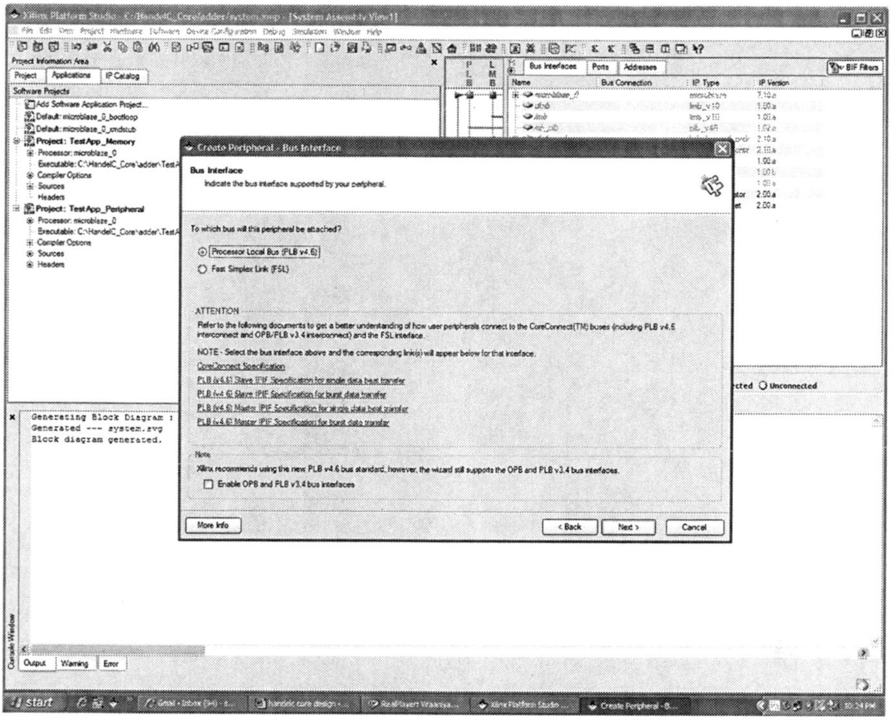

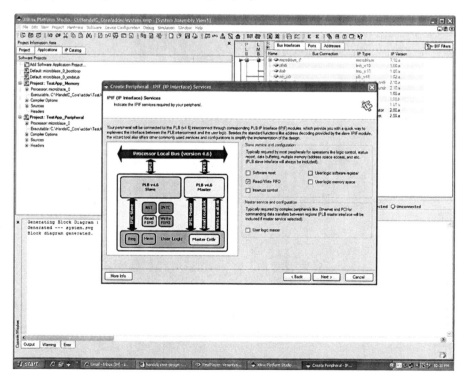

Fig. 5.15 Selecting the bus for connecting the peripheral to Micorblaze

Fig. 5.16 In this application w have selected FIFO for buffering the data

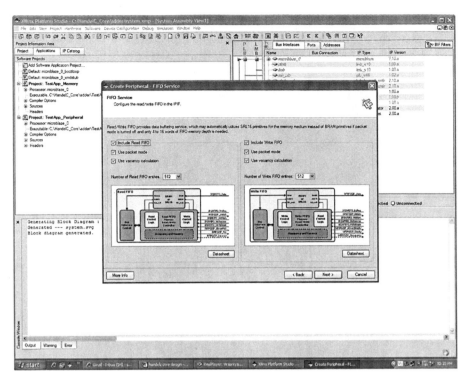

Fig. 5.17 FIFO size specified as 512 bytes + 512 bytes as read and write

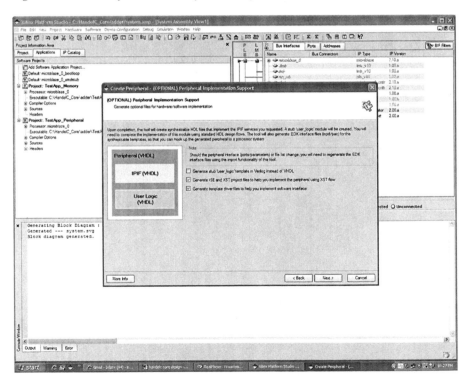

Fig. 5.18 Generating XST and ISE for implementing peripherals using XST flow. Template drivers are also selected here

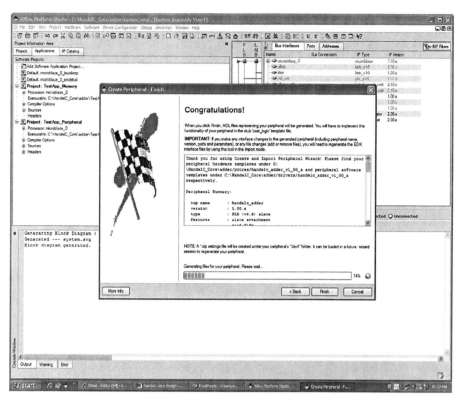

Fig. 5.19 With this step the blank core is created

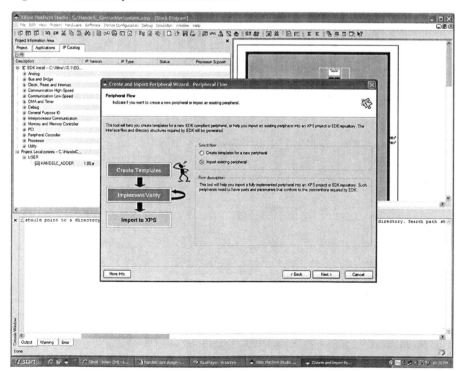

Fig. 5.20 The created peripheral is imported to the project

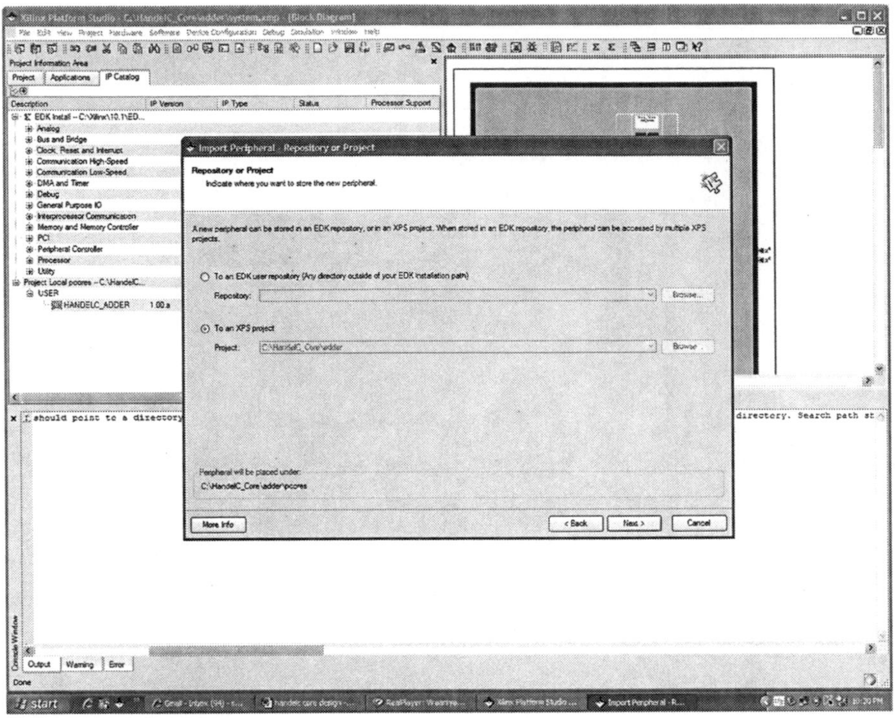

Fig. 5.21 Specifying the project for linking the created peripheral core

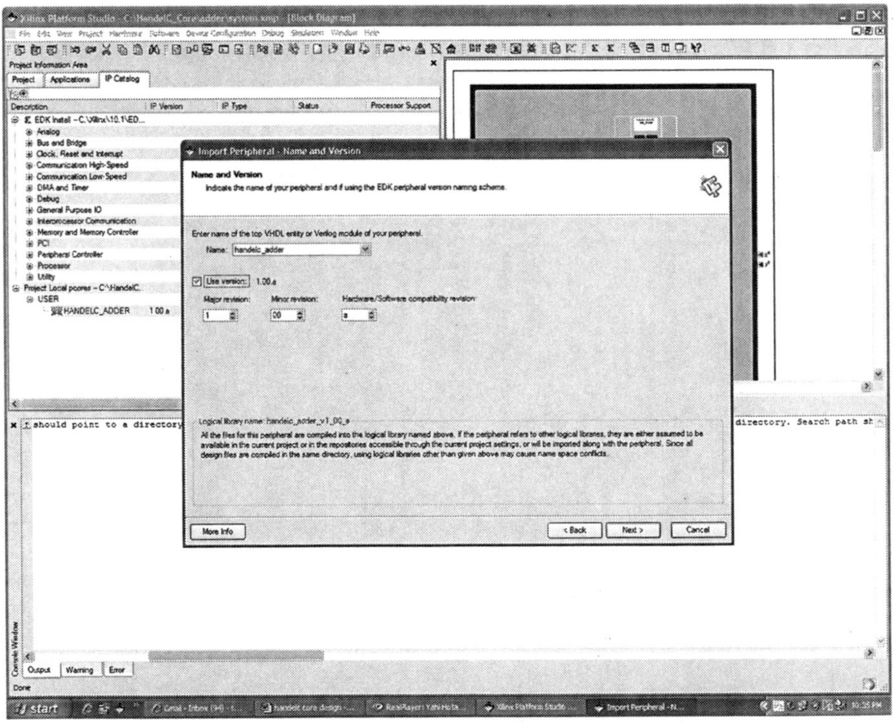

Fig. 5.22 Selecting the previously generated core with due attention given to the revision information

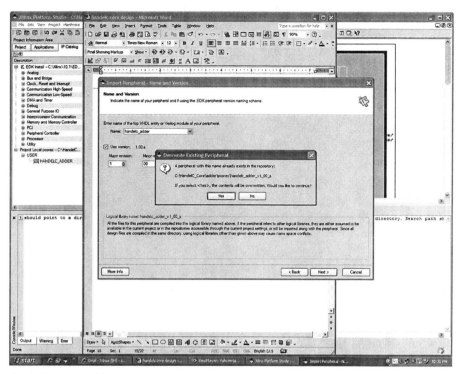

Fig. 5.23 Overwriting the core over the existing core

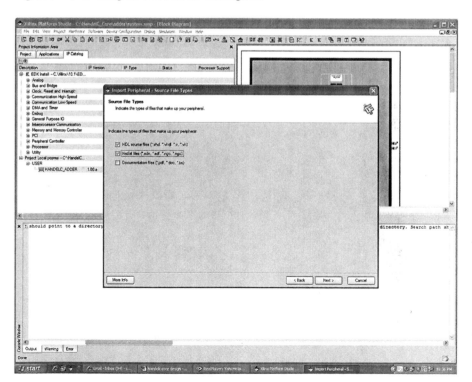

Fig. 5.24 Selecting the source file. In this case we have selected the EDIF generated by the Handel C adder core. Here the EDIF automatically maps into the VHDL for the compatibility with Microblaze

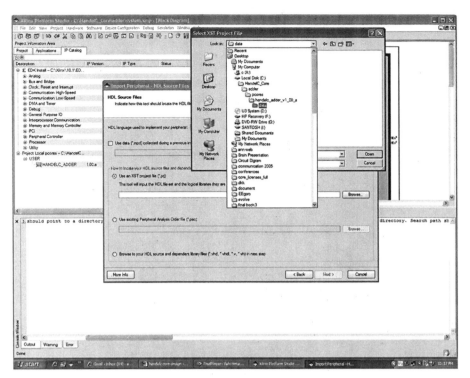

Fig. 5.25 Specifying the path for saving the VHDL file

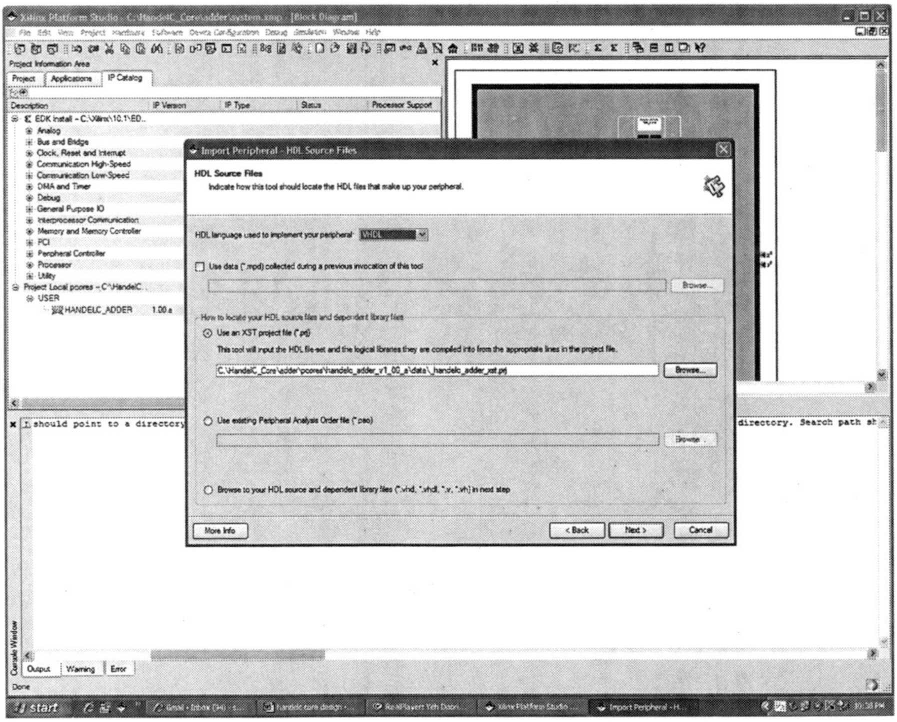

Fig. 5.26 Selecting the.XST file of the core by adopting proper path

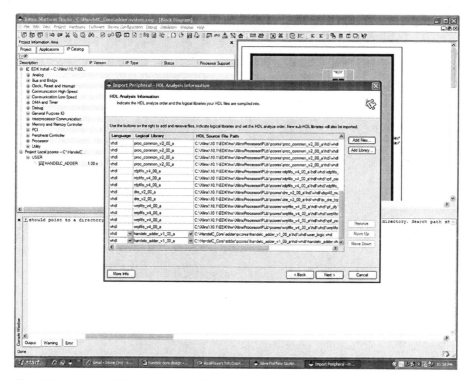

Fig. 5.27 Here the EDK shows successful creation of two VHDL files the first one is in VHDL pertaining to core and the second one is the user logic which is also in VHDL

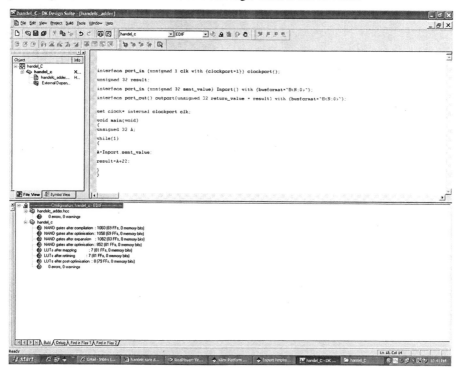

Fig. 5.28 Generating the EDIF of the Handel C Adder core

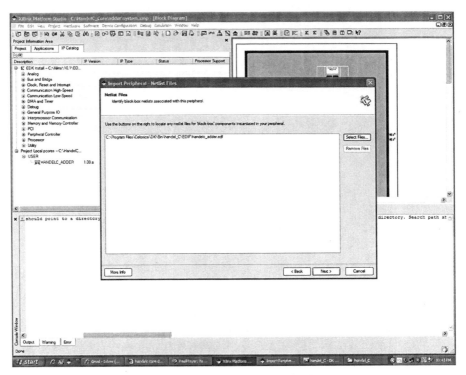

Fig. 5.29 The EDK seeks path for the netlist of the 'Adder Core'

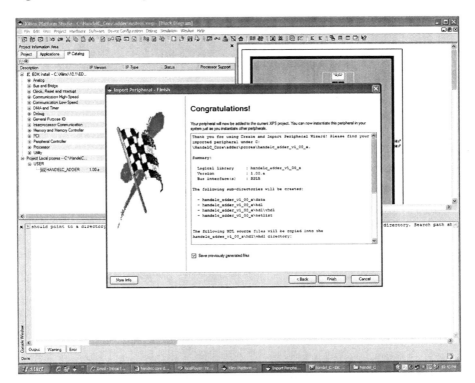

Fig. 5.30 EDK showing the successful generation of the 'Adder Core' with other information such as version etc

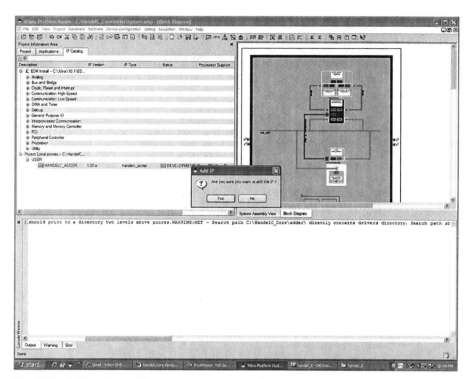

Fig. 5.31 Adding 'Adder Core' to the main project of Microblaze

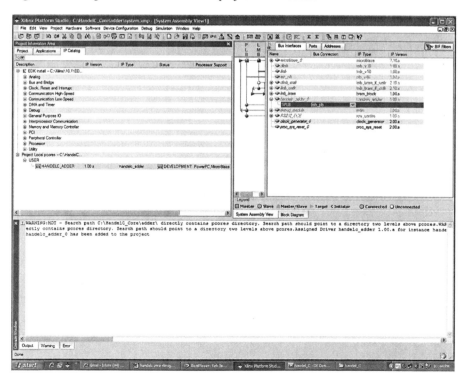

Fig. 5.32 Selecting the PLB bus interface for the 'Adder Core' to facilitate its communication with Microblaze

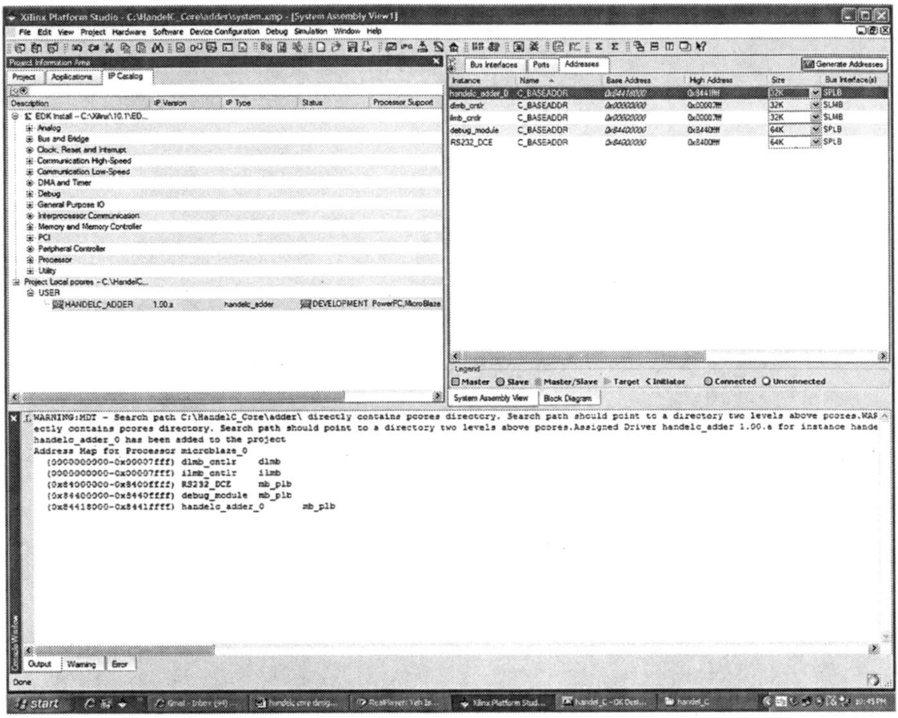

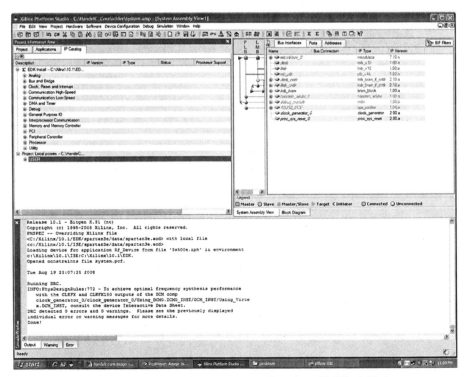

Fig. 5.33 Deciding the Memory Map for the System

Fig. 5.34 Generating the bitstream of the total core (Microblaze+UART+ Adder)

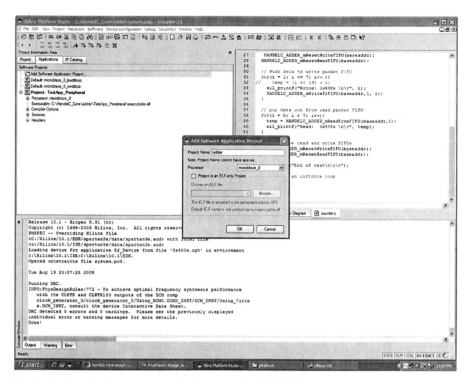

Fig. 5.35 Creating the software driver for the adder in ANSI C

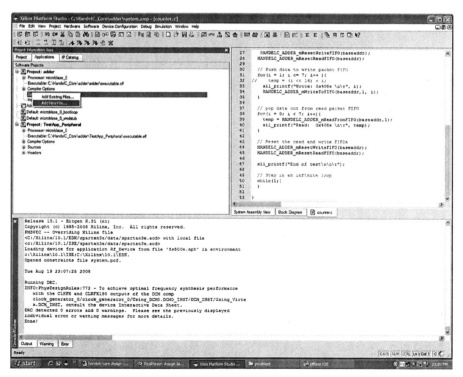

Fig. 5.36 Adding files to the Software part of the project

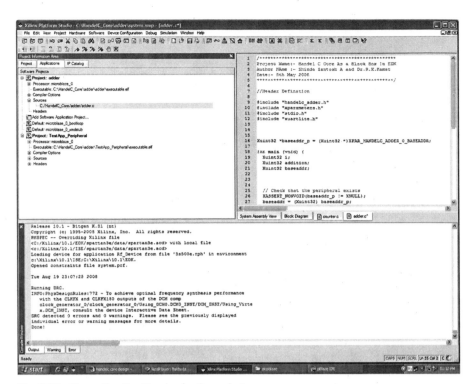

Fig. 5.37 Generating the libraries for the project

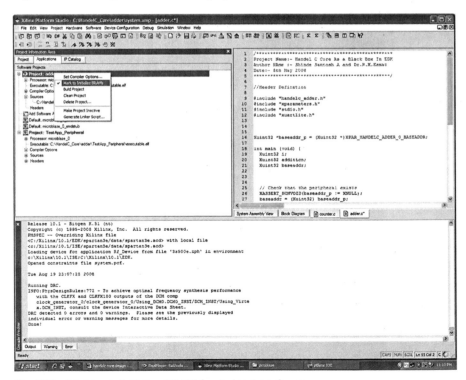

Fig. 5.38 Marking to initialize BRAM for storing the software

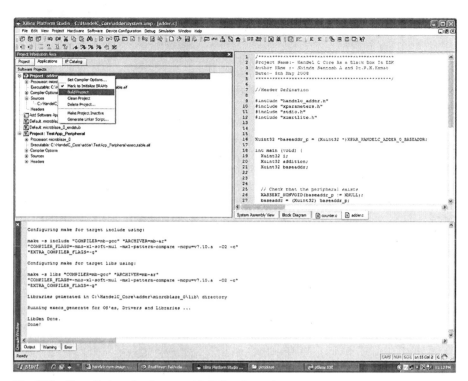

Fig. 5.39 Compilation of software and building the project

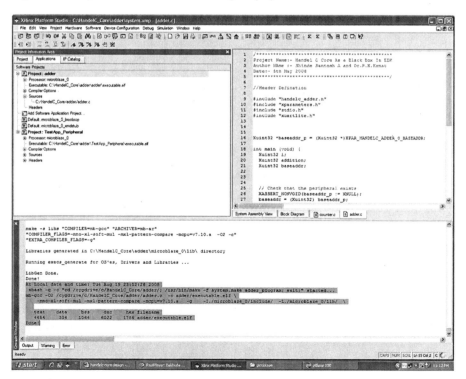

Fig. 5.40 The SoC comprising of Hardware and Software is successfully built

Here we have chosen the Microblaze as we are using the 'Xilinx Spartan 3E Starter Kit'. MicroBlaze is a 32 bit soft processor core having RISC architecture. It has 2–64 KB instruction and data caches along with the Barrel Shifter, Hardware multiply and divide and OPB and LMB bus interfaces. The chase selection can be done as per the need through EDK.

The selection of the processor and board follows by the selection of the peripherals. As shown in Fig. 5.6, the UART is selected. Its parameters are configured as baud rate of 9600 BPS for facilitating communication with the PC hyperterminal. As shown in Fig. 5.7, the UART is configured as standard input/output to acquire the data from the Microblaze and pass it to the PC hyperterminal. Rest of the steps and their discussion follows as the tile of the diagrams itself. The files generated while realizing the SoC are as shown in Table 5.1. The realization is also a classic example of hardware – software co-design in a SoC paradigm as the software drivers are built in ANSI C.

Program 5.6: Handel C code for adder
```
********************************************************************
interface port_in (unsigned 1 clk with {clockport=1}) clockport();
unsigned 32 result;
interface     port_in     (unsigned     32     sent_value)     Inport()     with
{busformat="B<N:0>"};
interface  port_out()  outport(unsigned  32  return_value  =  result)  with
{busformat="B<N:0>"};
set clock= internal clockport.clk;
void main(void)
{
unsigned 32 A;
while(1)
{
A=Inport.sent_value;
result=A+22;
}
}
********************************************************************
```

Program 5.7: Handel C core as a black box in EDK
```
********************************************************************
Project Name:- Handel C Core As a Black Box In EDK
Author NAme :- Shinde Santosh A and Dr.R.K.Kamat
Date:- 5th May 2008
//Header Definition
#include "handelc_adder.h"
#include "xparameters.h"
#include "stdio.h"
#include "xuartlite.h"
```

```c
Xuint32 *baseaddr_adder = (Xuint32 *)XPAR_HANDELC_ADDER_0_BASE-
ADDR;
int main (void) {
Xuint32 i;
Xuint32 addition;
Xuint32 baseaddr;
// Check that the peripheral exists
XASSERT_NONVOID(baseaddr_adder != XNULL);
baseaddr = (Xuint32) baseaddr_adder;
while(1)  //infinite Loop
{
//Reser FIFO
HANDELC_ADDER_mResetWriteFIFO(baseaddr);
HANDELC_ADDER_mResetReadFIFO(baseaddr);
// PUSH data to write packet FIFO for Addition
for(i = 1; i <= 7; i++)
{
xil_printf("Sent Value For Adder: 0x%08x \n", i);
HANDELC_ADDER_mWriteToFIFO(baseaddr,1, i);
}
// POP Adder Result out from read packet FIFO
for(i = 0; i < 7; i++){
addition = HANDELC_ADDER_mReadFromFIFO(baseaddr,1);
xil_printf("Result Of Addition: 0x%08x \n", addition);
}
// Reset FIFOs
HANDELC_ADDER_mResetWriteFIFO(baseaddr);
HANDELC_ADDER_mResetReadFIFO(baseaddr);
}
}
```
****

Chapter 6
Rapid Prototyping of the Soft IP Cores on FPGA

6.1 Prototyping Philosophy

According to the SIA Technology Roadmap, by the end of this decade the semi-conductor industry will manufacture chips with four billion transistors, thousands of pins, and clock speeds of 10 GHz. In order to increase design productivity, a new design flow has recently emerged, based on the reuse of portable IP cores. An IP core is a block of logic or data that is used with a field programmable gate array (FPGA) or an application specific integrated circuit (ASIC). The increasing gap between design productivity and chip complexity, and emerging System-on-Chip (SoC) have led to the wide utilization of reusable intellectual property (IP) cores [79]. For an SoC chip, the validation of hardware, software, and firmware on a common platform can be accomplished using FPGA-based prototypes. FPGA prototypes make it possible for SoC designs to be delivered on time, on budget, and on market target [77]. The reward of FPGA based prototyping is getting better designs sooner, in which hardware and software components are integrated before final silicon. Today, the electronic system design community is mainly concerned with defining efficient System-on-Chip (SoC) design methodologies in order to benefit from the high integration capabilities of current FPGA technologies on the one hand, and manage the increasing algorithmic complexity of applications on the other hand. Rapid prototyping is considered as a key to speed up the system design [78]. The popularity of the FPGA based prototyping is evident from the survey commissioned by Synplicity® Inc. in December 2004, wherein more than 20,000 developers around the world were questioned as to their hardware-assisted ASIC verification strategy. The results showed that 1/3 of today's ASIC designs are veri-fied by means of an FPGA-based prototype [80]. Thus, it is becoming increasingly necessary to create prototypes of ASIC designs that run 'at speed' in the context of the system. The most cost effective technique emerged out to achieve this level of performance is to create an FPGA-based prototype.

This chapter is dedicated to the theme of FPFA based rapid prototyping of soft IP cores developed in Handel C. The core theme is supported by development of three case studies.

R.K. Kamat et al., *Unleash the System on Chip using FPGAs and Handel C,*
DOI 10.1007/978-1-4020-9361-6_6, © Springer Science+Business Media B.V. 2009

6.2 Rapid Design of Fuzzy Controller Using Handel C

This case study reports a novel design of fast fuzzy processor using Handel C. A case study of the model of a fuzzy processor for optimizing the temperature of water bath is developed and the system has been successfully validated on Xilinx Spartan III FPGA.

6.2.1 Cramming Software Centric Fuzzy Logic on the SoC

Fuzzy logic was first developed by Zadeh [81] in the mid-1960 for representing uncertain and imprecise knowledge. The fuzzy logic techniques have been successfully applied in a number of applications such as computer vision, decision making, and system design including ANN training. However, the most extensive applications are in the area of control systems where examples include controllers for cement kilns, braking systems, elevators, washing machines, hot water heaters, air conditioners, video cameras, rice cookers and photocopiers [82]. It is estimated that the number of industrial and commercial applications will approximately double every year. It is estimated that the demand for fuzzy hardware specifically fast fuzzy processors will increase significantly and will be one of the most interesting businesses in the near future [83]. The main performance bottleneck in realizing the hardware for fuzzy processor is the inherent nature of the fuzzy programs vis-à-vis their hardware paradigms available for realization such as VHDL. The former is software centric with long chain of rule base while the later emphasizes on the concurrency and parallelism. The fuzzy programs are more of IF-THEN-ELSE type rule base which the software professionals find it difficult to obtain their hardware counterpart. In this application, a novel framework for rapid development and synthesis of fuzzy controller is presented. The methodology is based on Handel – C [84] with the output EDIF tested on Xilinx Spartan III FPGA platform. Design flow with mixed tool sets has been developed in order to overcome the inherent limitations of the individual tools. The design problem taken up is development of fuzzy controller for water bath temperature. Focus is on the realization of the fuzzy rule base into hardwired logic by expressing the same in Handel –C paradigm and finally transforming the same to a SoC paradigm.

6.2.2 Design Problem

The purpose of the developed fuzzy controller is controlling the temperature of the water and vary it so as to enhance the comfort level of the person taking bath. For instance, in a chilly winter, the water will be too cold; the hot water valve will be opened more to add hot water. The rule base for the system is as follows:

Table 6.1 Rule evaluation table for the proposed controller

IF				THEN
Rule	Season	Logical operator	Temperature of water	Control valves for to adjust the proportion of hot and cold water
1	A_1	AND	B_1	C_1
2	A_1	AND	B_2	C_1
3	A_1	AND	B_3	C_2
4	A_1	AND	B_4	C_3
5	A_2	AND	B_1	C_1
6	A_2	AND	B_2	C_2
7	A_2	AND	B_3	C_3
8	A_2	AND	B_4	C_3

Notations used in the table are as follows: Season: Winter = A_1, summer = A_2; Temperature of the water: Cool = B_1, Warm = B_2, Hot = B_3, Very hot = B_4; Controlling Valves of hot and cold water: Increase = C_1, Maintain = C_2, Decrease = C_3.

1. **If** season is winter **and** water is Cool **then** increase the proportion of hot water.
2. **If** season is winter **and** water is warm **then** increase the proportion of hot water.
3. **If** season is winter **and** water is Hot **then** maintain proportion of hot and cold water.
4. **If** season is winter **and** water is Very Hot **then** Decrease proportion of hot water.
5. **If** season is summer **and** water is Cool **then** Increase proportion of hot water.
6. **If** season is summer **and** water is Warm **then** maintain proportion of hot and cold water.
7. **If** season is summer **and** water is Hot **then** decrease proportion of hot water.
8. **If** season is summer **and** water is Very Hot **then** decrease proportion of hot water.

Season, Temperature of water and Control have been notified by the notations A, B and C respectively in order to derive the rule evaluation table for the proposed controller.

6.2.3   Fuzzification and Defuzzification

Trapezoidal membership functions are used for fuzzification and expressed in-terms of two slopes and points as shown in Fig. 6.1. Y axis defines degree of the membership (μ) expressed in the value interval between 0 and 1. X axis defines the universe of discourse and is divided into three segments 1, 2 and 3. The degree of the membership depends on the location of the input value of the season temperature. The season is further defined by the values in the particular segment. For example, the season winter is defined when the ambient temperature is less than the 20°C and summer is defined above 25°C. Calculation of the degree of the membership values in different segments is as follows. For segment 1 the slope is growing left to right

Fig. 6.1   Defining
membership function
for different seasons

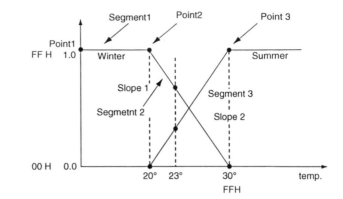

and here μ is limited to maximum value 1. For the segment 2 the slope is downward hence limited to minimum value 0. Refer equation 'I' for the formula of μ pertaining to segment 2.

$$\mu = 1 - (\text{input value - point 2}) * \text{slope1} \text{ ------------(I)}$$

For segment 3 the slope is upward form left to right and the μ is limited to the maximum value 1 as per the mathematical expression 'II'

$$\mu = (\text{input value - point 1}) * \text{slope2} \text{ ------------(II)}$$

The sample calculation of the membership function is worked out here as an example. Assuming 8 bit resolution μ = 1 equals to the FFH or 255 decimal. The slope is calculated as

$$\text{Slope 1} = 1/(30-20) = \text{FF H}/(\text{FFH-A8 H}) = 02 \text{ H ------------(III)}$$

Similarly Slope 2 = 02 H
If the season is winter and the temperature is 23°C i.e. (BE H), it lies in the segment 2 which corresponds to
$$\mu = 1 - (\text{input value - point 2}) * \text{slope1}$$

Table 6.2   Range of water temperature expressed in terms of Fuzzy variable

Fuzzy variable	Range (°C)
Cool	0–30
Warm	20–40
Hot	30–60
Very hot	45–100

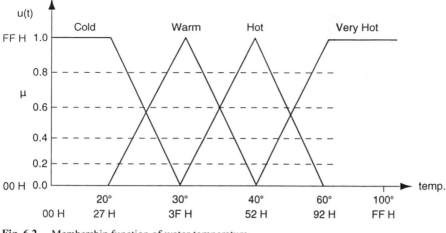

Fig. 6.2 Membership function of water temperature

i.e. μ= FF H – (BE H – A8 H) * 02 H = D3 H ------------(III)

Table 6.2 shows the range of the water temperature expressed in-terms of fuzzy variable.

The membership function of temperature of water is shown in Fig. 6.2.

6.2.4 Design Flow for Prototyping the Fuzzy Controller

The design flow adopted is shown in Fig. 6.3.

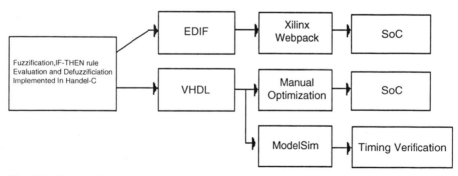

Fig. 6.3 Design flow

6.2.5   Handel C Implementation

The rule base expressed in Handel –C is given below. Similar routines have been developed for other seasons.

Program 6.1: Handel C code for fuzzy controller realization
```
*************************************************************
Fuzzy Controller Program
set clock = external "p77";
void main (void)
{
unsigned 8 temp_season,flow_temp;
unsigned 2 volve;
interface bus_in(unsigned 8 in2) inp1()with {data = {"p15", "p16", "p17",
"p18","p21","p22","p23","p24"}};
interface bus_in(unsigned 8 in1) inp() with {data = {"p3", "p4","p5",
"p7","p9","p10","p11","p13"}};
interface bus_out() out(volve)with {data = {"p25","p26"}};
while(1)
{
temp_season = inp.in1;
flow_temp = inp1.in2;
if(temp_season>=0x80)
{
//season=summer
if (flow_temp<=0x1E && flow_temp>=00)
{
volve=0b10; //increase temp
}
else if (flow_temp<=0x28 && flow_temp>=0x15)
{
volve=0b11; //maintain temp
}
else if (flow_temp<=0x3C && flow_temp>=0x1E)
{
volve=0b01; //decrese temp
}
else
{
volve=0b01;
}
}
else
{
//season = winter
if (flow_temp<=0x1E && flow_temp>=00)
{
```

```
volve=0b10;
}
else if (flow_temp<=0x28 && flow_temp>=0x15)
{
volve=0b10;
}
else if (flow_temp<=0x3C && flow_temp>=0x1E)
{
volve=0b11;
}
else
{
volve=0b01;
}
}
}
}
```

Number of errors: 0
Number of warnings: 2
Logic Utilization:
Number of Slice Flip Flops: 26 out of 7,168 1%
Number of 4 input LUTs: 32 out of 7,168 1%
Logic Distribution:
Number of occupied Slices: 26 out of 3,584 1%
Number of Slices containing only related logic: 26 out of 26 100%
Number of Slices containing unrelated logic: 0 out of 26 0%

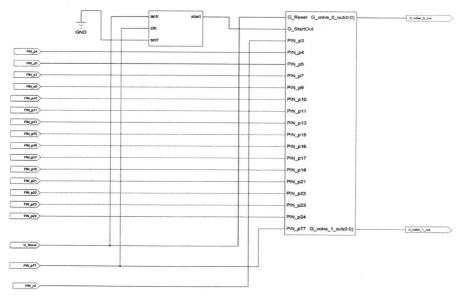

Fig. 6.4 RTL realization

Total Number of 4 input LUTs: 32 out of 7,168 1%
Number of bonded IOBs: 13 out of 141 9%
IOB Flip Flops: 11
Number of GCLKs: 1 out of 8 12%
****

6.2.6   Co Simulation with ModelSim

Co-simulation provides a framework for co-operation between two separate distinct
tools from independent vendors to pave the benefits of each of them in the SoC
design. Timing analysis for checking the realization of the constraints is decisive in
SoC and cosimulation addresses these woes. Moreover design platforms like Handel C, though improve design productivity, their analytical marriage with the well
proven and time tested design platforms like VHDL gives a chance of optimization
in a mixed design environment.

Fuzzy logic controller SoC designed by adopting the mixed design flow is as
shown in Fig. 6.3. The EDIF generated by the Handel C compiler is ported to the
Xilinx Web pack Version 9.2 for synthesizing the SoC on the Spartan III FPGA.
At the same time, the SoC is also been synthesized by converting the Handel C

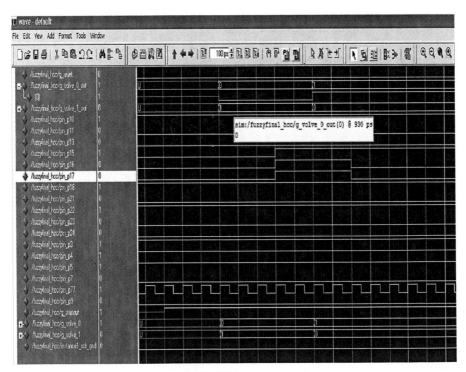

Fig. 6.5 Timing verification using the ModelSim

code into VHDL and by adopting the manual optimization. The RTL view of the realization is as shown in Fig. 6.3. The third and important phase is the cosimulation by porting the VHDL code to the ModelSim environment that gives valuable timing verification. A screenshot of the cosimulation using the ModelSim is shown in Fig. 6.6.

6.2.7 SoC Prototyping and Final Device Specifications

The system was successfully validated on Xilinx Spartan III FPGA. The novel framework presented in this section facilitates the rapid conversion of the fuzzy logic algorithms expressed in any ANSI-C like language directly in terms of hardware. As the Handel-C supports a large set of ANSI-C constructs, easy porting between two languages is possible. Moreover, the above approach facilitates easy debugging and provides good scalability. The usage of the third party tool like Modelsim helps in extensive testing in less time. Implementation on the FPGAs provides relatively quick implementation from concept to physical realization as compared to the ASIC or custom IC implementations. Moreover, it reduces the initial cost, time delay and inherent risk of a conventional masked gate array. Thus the presented approach provides a fast route for hardware prototyping of the fuzzy logic algorithms and realizing them in terms of semicustom ASICs with high efficiency with rapid development cycle.

6.3 Packet Processor Core for Inculcating Embedded Network Security using Mixed Design Flow

This case study presents the design and development of the 'Embedded Security Core' based on the TCP/IP packet processing. The design objective is to realize a tailor made TCP/IP packet processor that can be integrated as and when required to any internet enabled embedded device. A mixed design flow consisting of the incorporation of tools from third party has been adopted for the simulation, testing, debugging and generation of the RTL model. The reported 'packet processor core' has numerous potential applications in passive monitoring of the networking setup, security appliances like firewalls, Network on Chip (NoC) devices, internet enabled smart appliances etc. It can also be used as a distributed network security appliance by enforcing it to key locations in the network infrastructure for instance, at the end user PC, at remote switches, routers and even in the data center.

6.3.1 From SoC to NoC

There is a general agreement in the design community that the near future will be dominated by the ubiquitous Internet enabled devices. Just one such prediction is that by 2010, 95% of Internet-connected devices will not be computers. So if they are not computers, what will they be? Embedded Internet devices [129]. With the growing consensus of Internet Enabled Smart Products, the concept of Network on Chip (NoC) is rapidly penetrating in the SoC arena.

Network-on-a-chip (NoC) is an emerging approach to System-on-a-chip (SoC) design. The obvious advantage of integrating the network functionality on the SoC is accommodation of multiple asynchronous clocking and facilitating the on-chip communication to achieve significant throughput improvements over conventional bus systems. Moreover adoption of the NoC methodology on Silicon paves the benefits such as increased parallelism due to simultaneous operation of the many links through packet processing and scalability as the internet enabled embedded products can be updated and upgraded on the fly.

The NoC on silicon lays it foundation on the well established field of computer networking, however with a difference. Due to the spatial and temporal constraints in the Silicon arena, there is a trend of tailor made packet processing and integration of customized protocol stacks. The NoC achieved bearing the above mentioned principle enables the well separation of the computational and network functionality and thus achieves the improved performance.

Security is a major issue in today's communication networks. The performance pressures on implementing effective network security monitoring are growing fiercely due to rising traffic rates, the need to perform much more sophisticated forms of analysis, the requirement for inline processing, and the collapse of Moore's law for sequential processing. Given these growing pressures, it is time to fundamentally rethink the nature of using hardware to support network security analysis [85]. Many researchers have adopted different strategies for implementing the network security. The traditional approach for the network security is installing the intrusion detection software (IDS) per host. However this leads to many drawbacks. With the software IDS, it is not only harder to correlate network traffic patterns that involve multiple computers but also poses a challenge in coping up with the heterogeneous environment having mixed machine configurations and operating systems. It is also not very difficult to disable the IDS by the attackers. The literature survey reveals that, there is a growing upcoming trend to use Field Programmable Gate Arrays (FPGA) based customized Network On Chip (NoC) and Offload the software processing to hardware realizations. But it turns out to be a costlier solution and therefore hardware-software based hybrid solutions for the security scenario are widely discussed in literature [86, 130].

6.3.2   A Novel Frame work for Designing NoC

FPGAs have manifested their presence in the network on chip paradigm due to their various attributes. The foremost reasons are their capability to suit the real time constraints, increasing dynamism to address the pertinent issues of latency and throughput and faster design and prototyping cycle that facilitates robust design from testability point of view. FPGA based NoCs have been reported widely in the literature [88–91]. However, few researchers have rightly reported that there is no 'one size fits all' NoC architecture [92], as different silicon systems have very different requirements from their NoCs viewpoint. This is especially true in an FPGA based environment where the design metrics such as space, time are very stringent [130]. In this application we are proposing a novel framework for designing of the FPGA based NoC. A specific case study of the TCP/IP packet processor for embedded security enforcement is taken up.

6.3.3 Developing the FSM

A finite state machine (FSM) is a behavioral model comprising of a finite number of states, transitions between those states, and actions. It is an abstract model based presentation of a machine with a primitive internal memory. The FSM model of the packet processor core with details of the state and their transitions is shown in Fig. 6.6.

6.3.4 Simulation and Mixed Mode Design Aspects

This case study exhibits integration of third party tool i.e. ANSI C with Handel C. A separate ANSI C routine is incorporated in the main Handel C program for displaying the packet attributes, calculating the packet statistics for out of sequence packet analysis and alerting the custom messages to the user as regards to the bad IP or Denial of Access. The simulation window is shown in Fig. 6.7.

6.3.5 Handel C Implementation

Handel C based implementation for realizing NoCs gives various benefits. The NoC can be expressed at very high level of abstraction leaving the worries of the actual structural implementation. The built in Platform Abstraction layer (PAL) with Handel C features readymade macros required for the design of the NoC. The required headers such as Ethernet, console, RAM, display can be declared and invoked as and when required in the main program. Thus the PAL based design with predefined functionality to access peripherals via APIs offers fast prototyping and narrow development cycle. The present case study of packet processor showcases all the benefits of using the Handel C for realizing the NoC. Complete program listing is given below:

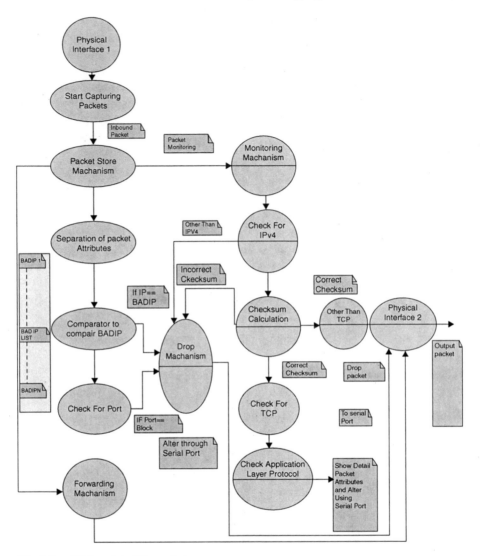

Fig. 6.6   FSM model of the packet processor core

Program   6.2 Handel C code for TCP/IP packet processor implementation
****

#define PAL_TARGET_CLOCK_RATE PAL_PREFERRED_VIDEO_CLOCK_
RATE
#include "pal_master.hch"
#include "pal_console.hch"
/*
 * Forward declarations
 */

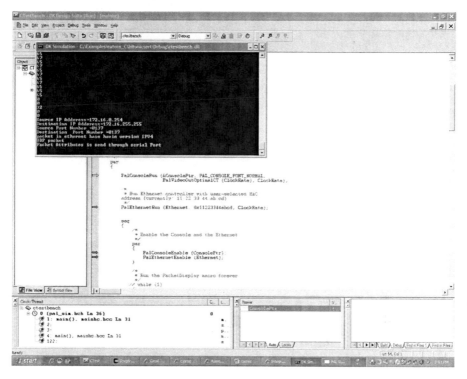

Fig 6.7   Simulation window of Packet Processor exhibiting the Mixed Mode Flow with ANSI C

static macro expr ClockRate = PAL_ACTUAL_CLOCK_RATE;

macro proc ReadPacket (ConsolePtr, Ethernet);
void DisplayByte (PalConsole *ConsolePtr, unsigned 8 Byte);
macro proc DisplayMAC (ConsolePtr, MACAddress);

macro proc PacketAnalyzer(value);
macro proc PacketMonitor(Value);

extern "C"
{
   void   DisplayData(unsigned char *Buffer, unsigned char Size);

  int printf(const char *format, ...);
}

chanout output with { outfile = "c:\\packet.dat" };

/*
 * Main program

```
*/
void main (void)
{
   macro expr Ethernet = PalEthernetCT (0);
   PalConsole *ConsolePtr;

//#define BUFFER_SIZE 1500
unsigned char BUFFER_SIZE;

unsigned   Data[40],d ;
unsigned i;
  /*
   * Check we've got everything we need
   */
  PalVersionRequire  (1, 2);
  PalVideoOutRequire (1);
  PalEthernetRequire (1);

  /*
   * Run Console and Ethernet in parallel with other code
   */
  par
  {

     PalConsoleRun (&ConsolePtr, PAL_CONSOLE_FONT_NORMAL,
             PalVideoOutOptimalCT (ClockRate), ClockRate);

     /*
      * Run Ethernet controller with user-selected MAC
      * address (currently: 11 22 33 33 ab cd)
      */
     PalEthernetRun (Ethernet, 0x11223333abcd, ClockRate);

     seq
     {
       /*
        * Enable the Console and the Ethernet
        */
       par
       {
         PalConsoleEnable (ConsolePtr);
         PalEthernetEnable (Ethernet);
       }

       /*
```

```
      * Run the PacketDisplay macro forever
      */
    // while (1)
      {
      par
                        {
                            ReadPacket (ConsolePtr, Ethernet);
            printf("Getting data from external C routine...\n");


                        }
        }
      }
    }
  }
```

```
/*******************************************************************
PacketDisplay (ConsolePtr, Ethernet)

******************************************************************/
macro proc ReadPacket (ConsolePtr, Ethernet)
{
   unsigned 1 Error;
   unsigned 16 Type;
   unsigned 48 Dst, Src;
   unsigned 11 Total_Length, Counter;
   unsigned 8 Data;
        unsigned char BUFFER_SIZE;
        unsigned  8 value[1500];

   /*
    * Strings to display on PAL Console
    */

   unsigned 8 ASCII;   /* variable to copy ASCII codes into for display */

   /*
    * Attempt to read ethernet packet
    */
   par
   {
      PalEthernetReadBegin (Ethernet, &Dst, &Src,
                  &Type, &Total_Length, &Error);
      Counter = 0;
   }
```

```
/*
 * If read was successful
 */
if (Error == 0)
{

    do
    {
      PalEthernetRead (Ethernet, &Data, &Error);
      if (Counter[3:0] == 0x0)
      {
        PalConsolePutChar (ConsolePtr, '\t');

      }
      else
      {
        delay;
      }

      PalConsolePutChar (ConsolePtr, ' ');
      par
      {
                            output ! Data;

                            BUFFER_SIZE=Total_Length[7:0];

                            value[Counter]=Data;

                            DisplayData(value, BUFFER_SIZE);

        if (Counter[3:0] == 0xf)
        {
          PalConsolePutChar (ConsolePtr, '\n');
        }
        else
        {
          delay;
        }
        Counter++;
      }
    }
    while (Counter != Total_Length);

        PalConsolePutChar (ConsolePtr, '\n');
```

```
                    PacketAnalyzer(value);

                    PacketMonitor(value);
              PalEthernetReadEnd (Ethernet, &Error);
      }
      else
      {
        delay;
      }
}




//************************************************************
macro proc PacketAnalyzer(value)

{
unsigned 8 type;
unsigned i;
static unsigned 8 badIP1[4]={172,16,0,1};
        static unsigned 8 badIP2[4]={172,16,0,1};
        static unsigned 8 badIP3[4]={172,16,0,2};
        static unsigned 8 badIP4[4]={172,16,0,102};
        static unsigned 8 badIP5[4]={172,16,0,3};

type=value[0];
if(type==0x45)
{
output!type;
printf("packet is Ethernet base having version IPV4\n");
//packet is ethernet
}
else
{
output!type;
printf("packet is not ethernet base\n");
}

type=value[9];
if(type==0x6)
{
printf("TCP Packet\n");
}
else if(type==0x11)
{
```

```
printf("UDP packet\n");
}
if(value[12]==badIP1[0] && value[13]==badIP1[1] && value[14]==badIP1[2]
&& value[15]==badIP1[3])
{
printf("Bad Packet\n");
}

else if(value[12]==badIP2[0] && value[13]==badIP2[1] && value[14]==badIP2[2]
&& value[15]==badIP2[3])
{
printf("Bad Packet\n");
}
}

macro proc PacketMonitor(value)
{
printf("Packet Attributes is send through serial Port\n\n");
}
****************************************************************
```

6.4 A Linear Congruential Generator (LCG) SoC

A linear congruential generator (LCG) represent one of the oldest and best-known pseudorandom number generator algorithms. The linear congruential generator (LCG) was proposed by Lehmer in 1948 [93]. The main advantage of the LCG method is speed efficiency owing to only a few operations per call. The Soft IP core of the random number generator in Handel C and its realization on the Xilinx Spartan III FPGA is given below:

Program 6.3: Handel C code for LCG core

```
****************************************************************
unsigned 32 X;
interface port_in (unsigned 1 clk with {clockport=1}) clockport();
interface port_in (unsigned 32 value1) Inport1() with {busformat="B<N:0>"};
interface port_in (unsigned 32 value2) Inport2() with {busformat="B<N:0>"};
interface port_in (unsigned 32 value3) Inport3() with {busformat="B<N:0>"};
interface  port_out()  outport(unsigned  32  return_value  =  X)  with
{busformat="B<N:0>"};
set clock= internal clockport.clk;
void main(void)
{
unsigned 32 A,C,M;
// take the value from user
```

```
A=Inport1.value1;
C=Inport2.value2;
M=Inport3.value3;
while(1)
{
X=(A * X + C)% M; //generate random number.and return a value to user.
}
}
```
****

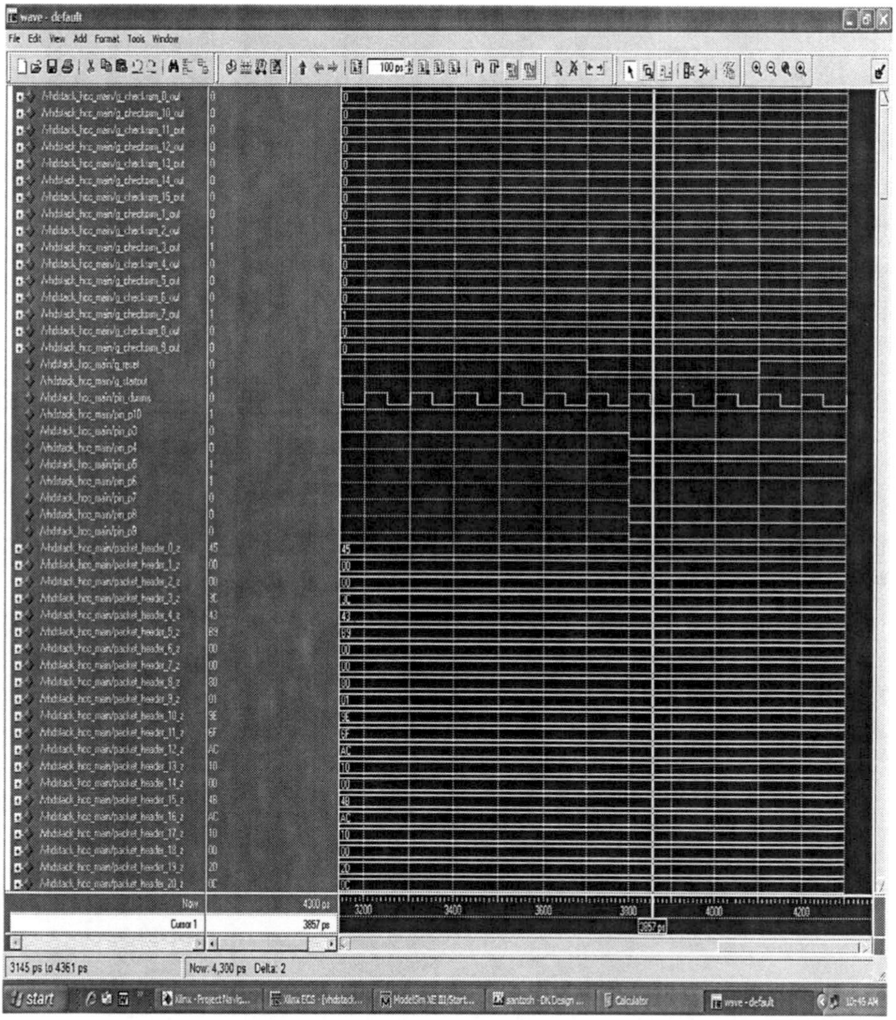

Fig. 6.8 Simulation report using ModelSim, the timeline (*yellow*) is the reference to validate the timing at a particular instance

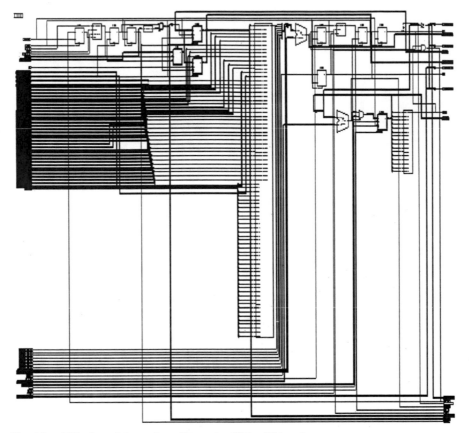

Fig. 6.9   RTL view of the packet splitter using Xilinx Webpack

6.4.1 Map Report Generated by Xilinx Webpack

Target Device: xc3s400
Target Package: pq208
Target Speed : -4
Mapper Version: spartan3 -- $Revision: 1.36 $
Mapped Date : Fri Aug 08 20:09:20 2008
Design Summary
Logic Utilization:

- Number of Slice Flip Flops   :   51 out of 7,168 1%
- Number of 4 input LUTs   :   3,054 out of 7,168
  42%

Logic Distribution:

- Number of occupied Slices   :   1,548 out of 3,584
  43%
- Number of Slices containing onlyrelated logic :   1,548 out of 1,548
  100%
- Number of Slices containing unrelated logic   :   0 out of 1,548 0%
- Total Number of 4 input LUTs   :   3,054 out of 7,168
  42%
- Number of bonded IOBs   :   130 out of 141 92%
- IOB Flip Flops   :   96
- Number of MULT18x18s   :   3 out of 16 18%
- Number of GCLKs   :   1 out of 8 12%
- Total equivalent gate count for design   :   40,832
- Additional JTAG gate count for IOBs   :   6,240

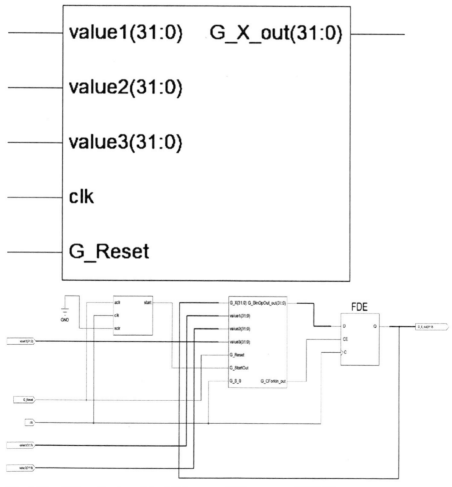

Fig. 6.10   RTL realization of the LCG core on Xilinx spartan III FPGA

- Peak Memory Usage   :   155 MB
- Total REAL time to MAP completion   :   9 s
- Total CPU time to MAP completion   :   7 s

6.5   Implementation of Reusable Soft IP Core of Blowfish Cipher

Blowfish is one of the fastest algorithms adopted by the large number of cipher suites and encryption products. Although used widely in desktop and laptop computers as a software implementation, the algorithm remained unexplored in the embedded system paradigm due to its large memory footprint. The present communication reports implementation of reusable Soft IP core of the 'Blowfish Cipher' especially designed from the viewpoint of memory and time constraints in the embedded domain.

6.5.1   Design Problem

Blowfish is a symmetric block cipher that is being used as a drop-in replacement for 'Data Encryption Standard' (DES) [118] or 'International Data Encryption Algorithm' (IDEA) [119]. It takes a variable-length key, from 32 to 448 bits, making it ideal for both domestic and exportable use. The algorithm was designed by Bruce Schneier in 1993, and immediately became famous owing to its fast execution speed, in comparison with the contemporary encryption algorithms. The noteworthy features of this algorithm includes its inclusion in public domain and thus ensures unpatented and license-free usage. However, inspite of the number of implementations in software, the literature survey reveals very few efficient hardware implementations [120], of this algorithm. The present application reports implementation of 'Blowfish' algorithm in Handel-C prototyped on the Xilinx Spartan x3s400 FPGA.

6.5.2   Implementation in Handel C

In this work, the 'Blowfish' cipher has been implemented in Handel-C, that uses much of the syntax of conventional C with the addition of inherent parallelism. In order to gain maximum benefit in performance from the target hardware the 'par' construct is used here. The 'macro' is built to get the advantage of 'spatial efficiency'. Complete listing of the program is as follows:

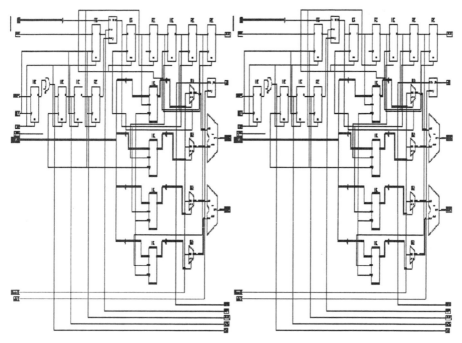

Fig. 6.11   RTL realization of the Blowfish Soft IP core

Program 6.4: Handel C code for blow fish implementation
****
//***** Program Listing of Blowfish Cipher Implementation *************??
```
unsigned 64 plaintext,Cipher_text,inbus;
unsigned 32 R_plaintext,L_plaintext,s_box,L_xor,R_xor;
unsigned 8 A1,A2,A3,A4;
unsigned i;
while(1)
{
plaintext=in1.inbus;
L_plaintext=plaintext[63:32];
R_plaintext=plaintext[31:0];
do
{
L_xor=L_plaintext^p_key[i];
A1=L_xor[31:24];
A2= L_xor[23:16];
A3=L_xor[15:8];
A4=L_xor[7:0];
//sbox value calculaion
s_box= (sbox0[A1] + sbox1[A2]) ^ (sbox2[A3] + sbox3[A4]);
R_xor=s_box ^ R_plaintext;
```

```
swap (L_xor, R_xor);
L_plaintext=L_xor;
R_plaintext=R_xor;
if(i==15)
{
break;
}
i++;
}
while(i<=15);
L_xor=L_plaintext^p_key[16];
R_xor=R_plaintext^p_key[17];
Cipher_text=L_xor@R_xor;
}
}
macro proc swap (L_xor, R_xor)
{
par
{
L_xor = R_xor;
R_xor = L_xor;
}
}
*****************************************************************
```

6.5.3   Prototyping on Spartan 3 FPGA

The program developed in Handel C is converted into the EDIF format for the synthesizing the hardware on the Xilinx Spartan 3 x3s400 FPGA for the target Package
pq208, having the target Speed -4. The synthesis view is shown in Fig. 6.1.

6.5.4   Design Summary

The design summary is as follows:
Design Summary for the Blowfish Implementation:
Logic Utilization:

Number of Slice Flip Flops	:	154 out of 7,168 2%		
Number of 4 input LUTs	:	150 out of 7,168 2%		

Logic Distribution:

Number of occupied Slices:	91 out of	3,584	2%
Number of Slices containing			

only related logic:	91 out of	91	100%
Number of Slices containing			
unrelated logic:	0 out of	91	0%
Total Number of 4			
input LUTs:	150 out of	7,168	2%
Number of bonded			
IOBs:	131 out of	141	92%
IOB Flip Flops:	128		
Number of GCLKs:	1 out of	8	12%

Total equivalent gate count for design: 3,159

Additional JTAG gate count for IOBs: 6,288

The average connection delay for the implementation is 1.278 nS. The average connection delay on the 10 worst nets is 3.941 nS which indicate the temporal efficiency of the implementation.

****

Chapter 7
Soft Processor Core for Accelerated Embedded Design

The synergy of FPGA and soft processor cores has the budding potential to allow the integration SoC into a single FPGA chip. Embedded engineers often fight with the confront of improving performance. Discrete processors a.k.a. hard processors pose the following most striking drawbacks when it comes to embedding them for a particular application:

- Fixed selection of peripherals, most of them remain unutilized for the given application.
- No possible customization in clock frequency, that drags the entire system slow or too fast and ends in power inefficiency.
- Less life time of the processor family
- Incompatibility interms of package size when a particular processor is being upgraded, posing difficulties in PCB designing.
- Speed and interface incompatibility when multiple heterogeneous processors are required to work as coprocessors.

Programmable logic has reached such a state of advancement in terms of speed and density that it became a truly attractive alternative to the above mentioned RISC and CISC processors. It can form a 'matrix' within which processing, peripherals, data path, and algorithms can be placed to create powerful, flexible, and upgradeable systems. It is now available in forms and sizes that range from the traditional use as glue logic up to structured ASIC replacements and even further [103].

As a result of all the above mentioned advancements, 'Soft Processor Cores' have emerged as an ultimate remedy for addressing the pitfalls of hard RISC and CISC processors. 'A soft-core processor is a hardware description language (HDL) model of a specific processor (CPU) that can be customized for a given application and synthesized for an ASIC or FPGA target.' The main advantages of the soft processor cores are as follows:

- Greater possibility of Design Reuse.
- Cost Effectiveness.
- Customization and Flexibility.
- Scalability.
- Possibility of Hardware-Software portioning at an early stage.

R.K. Kamat et al., *Unleash the System on Chip using FPGAs and Handel C*, 141
DOI 10.1007/978-1-4020-9362-3_7, © Springer Science+Business Media B.V. 2009

- Platform independence.
- Greater invulnerability to obsolescence.
- Design Modularization.
- Advantage of the earlier design case studies.

7.1 Building SoC for Temperature Control Application Using Picoblaze

7.1.1 Design Problem

The theme of the present case study is rapid prototyping of the microcontroller based control systems by integration of the Soft IP core on FPGA. Soft IP core of Xilinx Picoblaze has been used on Spartan III FPGA. A case study of single setpoint temperature controller is taken up. Complete methodology and listing of the VHDL and assembly language program are given in the paper so as to facilitate the validation the results. Usage of resources and redundant hardware presented in the design summary guides the user to select an appropriate FPGA for the intended application. The results serve as a guiding source for implementation of full custom ASIC implementation with an added advantage of thorough testing in an accelerated time to market design paradigm.

7.1.2 Essence of Prototyping

Microcontroller has become the soul of the state of art instrumentation systems owing to the intelligence and flexibility achieved with its inclusion. Embedded microcontroller modules offer many advantages over the standard PC such as low cost, small size, low power consumption, direct access to hardware, and if available, access to an efficient preemptive real-time multitasking kernel. However, typical difficulties associated with an embedded solution include long development times, limited memory resources, and restricted memory management capabilities [104]. In order to make the microcontroller with the intended instrumentation, there is a growing trend of sporadically incorporation of system instrumentation features into processor architectures [105]. However, the design theme of microcontroller based mainly software centric implementations do not guarantee the improvement in performance and power efficiency in all the circumstances due to the general purpose architecture of the target microcontroller itself. The development of Field Programmable Gate Arrays has provided yet another opportunity to the instrumentation professionals to fine tune the controller architecture as per the process requirements. The FPGA based development platform enables the developer to test and add features in parallel without the need for repeating the complete testing of the designed instrument in an iterative fashion [106]. Thus the FPGA based compilation of the instrumentation and control application results in a highly optimized silicon implementation that provides true parallel processing with the performance and reliability benefits of dedicated hardware circuitry.

The advantages of the FPGA based design paradigm can be further enhanced combining with the soft processor cores developed by third party. These soft IP cores not only allow the integration of system design into a single FPGA device but also enables optimized division of the system functionality into hardware and software. Thus the FPGA platforms with soft IP cores are emerging for process control applications enabling the designer's customization of the processor core in an rapid system development schedules leading to less time to market. The present communication reports a case study of picoblaze based soft IP core integration on Xilinx Spartan III FPGA for single setpoint temperature control application.

7.1.3 Overview of Soft IP Processor Cores by Different Vendors

It is worthwhile here to present a brief review of the soft IP cores available from different vendors. Owing to the advantages of the soft IP processors cores towards reconfigurability, customization and emulation, many manufacturers have developed these cores either for a specific FPGA family or as a third party solution. Altera has developed ARM as a hard process core while NIOS and NIOS II as soft IP cores [107]. Similarly ATMEL and Quick Logic [108] have come out with AVR and MIPS as hard processor cores respectively. However, the most popular amongst the design community is Xilinx's PicoBlaze a fully embedded 8-bit RISC microcontroller core. The main features of Pico Blaze are its compactness, and cost-effectiveness (as it is provided as a free), and well documented VHDL source file with royalty-free re-use within Xilinx FPGA's. The VHDL listing of the PicoBlaze frees it from obsolescence as the same can be updated as per the design requirement. The PicoBlaze core can be embedded within the target FPGA, however with certain restriction to the device family selection.

The PicoBlaze [109] design was originally named KCPSM which stands for 'Constant(K) Coded Programmable State Machine' (formerly 'Ken Chapman's PSM'). Ken Chapman was the Xilinx systems designer who devised and implemented the microcontroller. The KCPSM3 core is designed for Spartan 3 family, whereas the KCPSM2 and KCPSM are meant for vertex 2 or vertex 2 pro Spartan II and Spartan 2(e) or vertex (E) respectively. In general the PicoBlaze soft processor provides 49 different instructions, sixteen 8-bit registers, 256 directly addressable ports, and a maskable interrupt. The program length is 256 instructions, and all address values are specified as 8bits contained within the instruction coding. The design is based on the RISC 'Harvard architecture' model with separate data and instruction ports. Its basic functionality is easily extended by connecting additional logic to the microcontroller's input and output ports. PicoBlaze delivers 50 million instructions per second (MIPS) much faster than commercially available microcontroller devices, yet occupies a tiny footprint of just 35 Configurable Logic Blocks (CLBs). This processor has an 8-bit bus, which means all registers and arithmetic operations are only 8 bits. There are 16 general purpose registers to quickly access data (i.e. S0 to SF), and programs can be up to 256 assembly language instructions long. The PicoBlaze has an internal memory for storing data (called the scratch-pad) with 64 locations. This is to store data which isn't used as frequently as other data. Operations can only be performed on data in registers, but there are only 16 registers. However, data can be stored and retrieved from the scratchpad to the reg-isters. There are two flags ZERO and Carry Flag. The ALU operation results affect the ZERO and CARRY flags. The PicoBlaze module has 256 input ports and 256 output ports. An 8-bit address value provided on the PORT_ID bus together with READ_STROBE or WRITE_STROBE signals indicates the accessed port. The port address can be either supplied in the program as an absolute value, or specified indi-rectly as the contents of any of the 16 registers.

7.1.4 Design Methodology

In this case study a temperature control application is developed by integration of the soft IP core of Picoblaze on Spartan III FPGA. The block schematic of the system is shown in Fig. 7.1. The ADC 0804 is interfaced in an handshake manner with the Picoblaze. e Current temperature and the set point is displayed on an 16 x2 LCD in an alternate manner. ON-OFF strategy of control is implemented with the assembly language instructions of the Picoblaze.

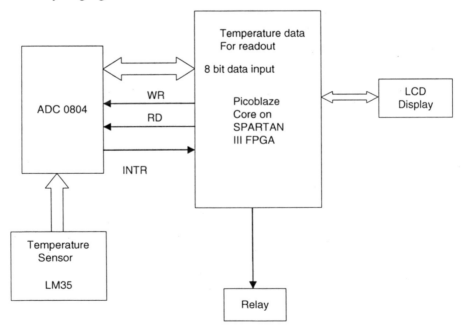

Fig. 7.1   Block diagram of the temperature control system application with integration of Picoblaze

7.1.4.1 Design Methodology

Design methodology followed for the implementation is given below in step by step manner.

1. PicoBlaze core was downloaded from the Xilinx website after registration.
2. The KCPSM3 version of the Picoblaze was downloaded as the target FPGA is SPARTAN III. The core comprises of VHDL files (named as KCPSM3), Program ROM, and an Assembler(KCPSM3.exe) and manual.
3. The top level VHDL program for the temperature controller was developed in VHDL. Behavioral model is developed the VHDL listing of which is given in the following point of the paper.
4. Appropriate assembly language program was developed for single setpoint temperature controller. The LCD display driver and the handshaking of ADC are the main parts of the program.
5. The assembly language program was executed using KCPSM3.exe. The output is in the form of ROM file which is nothing but the program developed for the Picoblaze.
7. The KCPSM3 VHDL file, generated ROM file are added to the project environment in Xilinx webpack.
8. The resulting file is then simulated, synthesized and the bit and MCS files are then dumped in the SPARTAN III FPGA board using JTAG port. This forms the customized core of the Picoblaze for the temperature controller application.

7.1.5 VHDL Program Listing

Program 7.1: VHDL code for temperature controller
```
********************************************************************
VHDL Listing of the Program
library IEEE;
use IEEE.STD_LOGIC_1164.ALL;
use IEEE.STD_LOGIC_ARITH.ALL;
use IEEE.STD_LOGIC_UNSIGNED.ALL;
entity temp_controller is
Port (ADC_in: in std_logic_vector(7 downto 0);
            relay_d : inout std_logic_vector(7 downto 0);
            lcd_d : inout std_logic_vector(7 downto 0);
            LCD_CONTROL : out std_logic_vector(3 downto 0);
            clk : in std_logic);
end temp_controller;
architecture Behavioral of temp_controller is
component KCPSM3
port (
address : out std_logic_vector(9 downto 0);
instruction : in std_logic_vector(17 downto 0);
```

```
port_id: out std_logic_vector(7 downto 0);
write_strobe: out std_logic;
out_port: out std_logic_vector(7 downto 0);
read_strobe: out std_logic;
in_port: in std_logic_vector(7 downto 0);
interrupt : in std_logic;
interrupt_ack: out std_logic;
reset : in std_logic;
clk: in std_logic
);
end component;
component control
port (
address : in std_logic_vector(9 downto 0);
instruction : out std_logic_vector(17 downto 0);
clk: in std_logic
);
end component;
--Processor signals
signal add_bus: std_logic_vector(9 downto 0);
signal inst_bus: std_logic_vector(17 downto 0);
signal port_add: std_logic_vector(7 downto 0);
signal data_out: std_logic_vector(7 downto 0);
signal data_in: std_logic_vector(7 downto 0);
signal write : std_logic;
signal read : std_logic;
begin
processor: kcpsm3
port map(
address => add_bus,
instruction => inst_bus,
port_id=> port_add,
write_strobe=> write,
out_port=> data_out,
read_strobe=> read,
in_port=> data_in,
interrupt => '0',
interrupt_ack=> open,
reset => '0',
clk=> clk
);
program: control
port map(
address => add_bus,
instruction => inst_bus,
clk=> clk
```

```
);
ADC_Register: process(write, data_in, port_add)
begin
if port_add= x"00" then
if read'event and read = '1' then
data_in<=ADC_in;
end if;
end if;
end process ADC_register;
LCD_Register: process(write, data_out, port_add)
begin
if port_add= x"04" then
if write'event and write = '1' then
lcd_d <=data_out;
end if;
end if;
end process LCD_register;
lcdcontrol_Register: process(write, data_in, port_add)
begin
if port_add= x"01" then
if write'event and write = '1' then
LCD_CONTROL<=data_out(3 downto 0);
end if;
end if;
end process lcdcontrol_register;
relay_Register: process(write, data_in, port_add)
begin
if port_add= x"08" then
if write'event and write = '1' then
relay_d<=data_out;
end if;
end if;
end process relay_register;
end Behavioral;
********************************************************************
```

7.1.6 Low Level Instruction to Enable the Application

Program 7.2: Assembly code for temperature controller
****
```
adc            DSIN    00
lcdcontrol     DSOUT 02
adccontrol     DSOUT 01
lcddata        DSOUT 04
```

```
relay           DSOUT 08
back: in s0,adc
      call display
      call delay
      jump back
display: load s1,$38
                call command
                load s1,$0e
                call command
                load s1,$01
                call command
                load s1,$80
                call command
                CALL CONVERT
                call data_disp
convert:load s4,$00
      load s5,$00
                load s1,s0
                        load s5,s1
                        load s2,$CC
convert1: sub s1,s2
                jump c, next
                add s4,$01
                jump convert1
                CALL BCD_CONVERT
                load s7,$00
                load s2,s1
                load s3,$0A
        next:sub s1,s3
                jump c, next
                add s7,$01
                CALL BCD_CONVERT
                RET
BCD_CONVERT:
                load sa,$00
                load s8,s1
                load s9,s8
                load s6,s8
AGAIN1:         SUB s6,$64
                load s0,s6
                jump c,NEXT1
                add sa,$01
                jump AGAIN1
NEXT1:
                load s6,S8
```

```
                    ADD s6,$64
                    load s5,s6
                    load s7,$00
                    ;CLR C
                    load S1,s5
AGAIN2:             SUB s1,$0A
                    load s5,s1
                    jump c, NEXT2
                    add s7,$01
                    jump AGAIN2
NEXT2:
                    ;CLR C
                    load s1,s5
                    ADD s1,$0A
                    ADD s1,$30
                    RET
command:
                    CALL READY
                    load S2,S1
                    load s3,$00
                    load s4,$01
                    out s3,lcdcontrol
                    out s4,lcdcontrol
                    out s3,lcdcontrol
                    RET
DATA_DISP:
                    CALL READY
                    load s2,s1
                    load s3,$04
                    out s3,lcdcontrol
                    load s4,$02
                    out s4,lcdcontrol
                    load s3,$01
                    out s3,lcdcontrol
                    load s3,00
                    out s3,lcdcontrol
                    RET
READY:
                    load s4,$00
                    load s5,$02
                    out s4,lcdcontrol
                    out s5,lcdcontrol
                    ;SETB P1.7
                    ;CLR P3.2
                    ;SETB P3.1
```

```
BACK1:
                    load s4,$00
                    load s5,$01
                    out s4,lcdcontrol
                    out s5,lcdcontrol
                    ;CLR P3.0
                    ;SETB P3.0
                    ;JB P1.7,BACK1
                    RET
                    END


delay:              load se,00
                    load sf,00
back3:              add se,$01
                    jump c, exit
back2:              add sf,$01
                    jump nc, back2
                    jump back3
exit:       ret
************************************************************************
```

7.1.7 Final SoC Specifications

The customized version of the Picoblaze for temperature controller application was successfully implemented on a Spartan III FPGA board. The same worked successfully under the umbrella of the assembly language program developed for this purpose. Moreover the design cycle in the course of simulation, verification presents valuable information given in the following paragraphs.

The design information as appeared in the Xilinx webpack is as follows:

```
Design Information
------------------
Command Line : C:/Xilinx/bin/nt/map.exe -intstyle ise -p xc3s400-pq208-4 -cm
area -pr b -k 4 -c 100 -tx off -o temp_controller_map.ncd temp_controller.ngd
temp_controller.pcf
Target Device  : x3s400
Target Package : pq208
Target Speed   : -4
Mapper Version : spartan3 -- $Revision: 1.16 $
Mapped Date    : Fri Oct 26 16:14:39 2007
```

Design summary gives the entire technical specifications of the customized Picoblaze implementation on the target Spartan III FPGA. The logical utilization is wound 1% with details as follows:

Logic Utilization:

- Number of Slice Flip Flops: 70 out of 7,168 1%
- Number of 4 input LUTs: 112 out of 7,168 1%

The logic distribution gives an idea of the extent to which the FPGA resources have been used.

Logic Distribution:

- Number of occupied Slices: 98 out of 3,584 i.e. 2%
- Number of Slices containing only related logic: 98 out of 98 i.e. 100%
- Number of Slices containing unrelated logic: 0 out of 98 i.e. 0%

The other finer details of the implementation are as follows:

- Total Number 4 input LUTs: 184 out of 7,168 2%
- Number used as logic: 112
- Number used as a route-thru: 4
- Number used for Dual Port RAMs   16

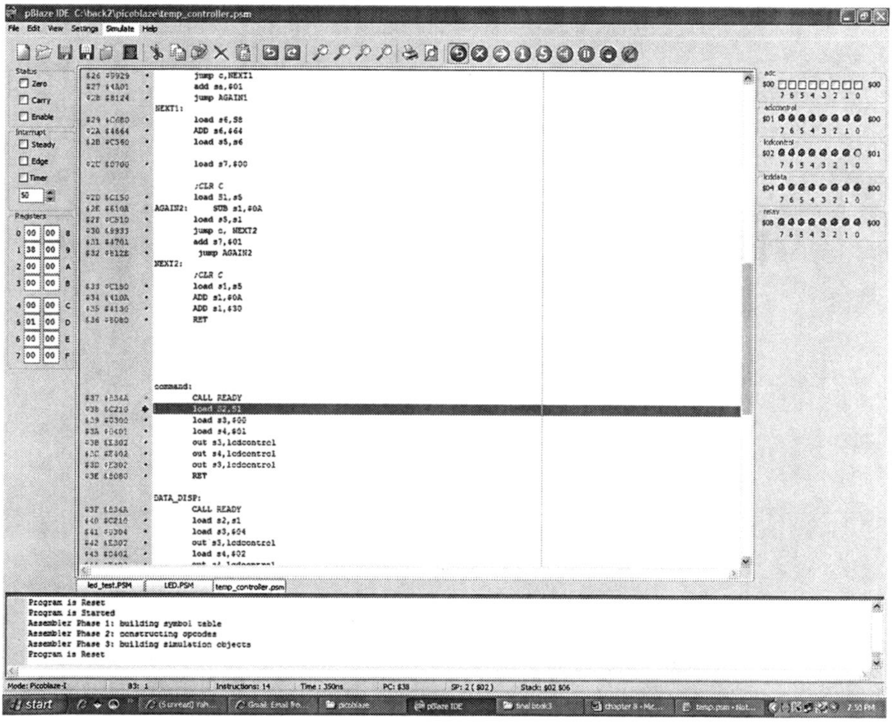

Fig. 7.2 PBlaze IDE interface

- (Two LUTs used per Dual Port RAM)
  Number used for 32x1 RAMs: 52
- (Two LUTs used per 32x1 RAM)
  Number of bonded IOBs: 29 out of 141 20%
- IOB Flip Flops: 28
- Number of Block RAMs: 1 out of 16 6%
- Number of GCLKs: 1 out of 8 12%

The total equivalent gate count used for the temperature controller design is
75,029. The analysis gives valuable results which guides the user as regards to the
effective selection of the target FPGA. It also presents a true picture of the used
resources and redundant hardware which could have been avoided in the following revision of the project to optimize the power and speed metrics of the system.
The RTL schematic shown in Figs. 7.2 and 7.3 are useful for the full custom ASIC
implementation for the dedicated temperature controller applications. Thus the Soft
IP core integrated with the FPGA offers a rapid prototyping environment to realize
the semicustom design and progress eventually towards the dedicated full custom
ASICs.

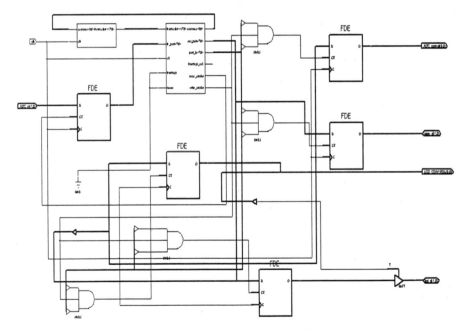

Fig. 7.3 LUT based implementation of the temperature controller

7.2 Hardware Software Codesign of SoC with Built in Position Algorithm

7.2.1 Design Statement

The paper communicates implementation of position algorithm of PID control using soft IP core on a FPGA platform. Xilinx picoblaze is used on Spartan III FPGA to implement the position algorithm for a temperature controller. Behavioral architecture of temperature controller is developed in VHDL with the control algorithm mapped to the software portion by means of the assembly language instructions of the processor. Usage of resources and redundant hardware presented in the design summary shows the superiority of the implementation as compared to the traditional control algorithms implemented on general purpose microcontrollers. The results serve as a guiding source for implementation of full custom ASIC implementation with an added advantage of thorough testing in an accelerated time to market design paradigm.

7.2.2 Building PID on Chip

Proportional-integral-derivative (PID) control is the most common and widely used process control algorithm commonly used in industries. The popularity of PID controllers is attributed to their effectiveness in a wide range of operating conditions, their functional simplicity and ease with which engineers can implement them using current computer technology [110]. Although many architectures exist for control systems, the PID controller is mature and well-understood by practitioners. For these reasons, it is often the first choice for new controller design [112]. With the evolution in the computing platforms; earlier in the form of microprocessors, later Programmable Logic Controllers and the latest towards the microcontrollers and Programmable Logic Devices there have been continues revisions and variations in the PID algorithm implementations. The implementations and review as regards to these evolving PID implementations is covered widely in the literature [111, 112]. There are several forms of digital PID implementations viz. interacting, parallel and gain independent. However, from the implementation in an microcontroller environment there are two widely used variations of PID algorithm namely displacement and velocity. In case of displacement PID algorithm, the control output is calculated in accordance with the displacement (position) of the process variable from the setpoint value (error term). In Here the control output is an absolute value. The velocity algorithm which is also called as incremental algorithm yields an output changing in each calculation period [113].

Literature survey regarding the digital PID implementations in a microcontroller paradigm reveals that the mathematical foundation of the algorithm is put forth so as to suit the instruction set of the state of the art microcontrollers. However, this leads to some bottlenecks in the performance as the target microcontroller architecture is not in tune with the PID implementation. This in turn leads to a typical situation of using only few resources of the microcontroller and thus affects the power and speed bottlenecks. In this communication a FPGA based hardwired implementation of PID algorithm is reported so as to customize the processor architecture as per the requirements. The implementation is centered around 'Picoblaze' the soft IP core by Xilinx Inc., which is integrated on Spartan III FPGA. A case study of temperature controller is taken up with implementation of position algorithm for the temperature control.

7.2.3   Picoblaze: A Soft IP Core from Xilinx

PicoBlaze is a 8-bit microcontroller developed and maintained by Xilinx and Ken Chapman. The microcontroller is described in VHDL and is to be implemented on Xilinx's different FPGAs and CPLDs. It is free to use as long as it is implemented in a FPGA or CPLD that comes from Xilinx [114]. An in-depth documentation of PicoBlaze along with the application notes are available online [115]. This is supplemented by online forum to help the users in case of any difficulty. A downloadable version of the PicoBlaze is available with all the related files including the VHDL definition of the processor, an assembler and the files that go with it [115]. The present application is developed with the latest version of the PicoBlaze [116] i.e. KCPSM3 owing to its features such as test and compare instructions, 64 byte scratch pad memory that works like an internal RAM etc. The KCPSM3 version of Picoblaze is implemented on Spartan III FPGA by Xilinx.

7.2.4   Design Problem and Methodology

7.2.4.1   Design Problem

The block schematic of the system is shown in Fig. 7.1. The ADC 0804 is interfaced in an handshake manner with the Picoblaze. Current temperature and the set point is displayed on an 16 x2 LCD in an alternate manner. A displacement / position type of PID algorithm is implemented by using the assemble language instructions of Picoblze. The processed digital output is applied through a DAC 0808 to a opto-coupled driver module comprising of triac and zero crossing detector. Power controlled output of the above mentioned module is applied to the heater unit for corrective action. The algorithm and other implementation issues are discussed in the following part of the paper.

7.2.4.2 Design Methodology

The position algorithm has been widely described in the literature, however the form useful from Picoblaze instruction set point of view is chosen here [116]. Pseudo code for the same is as follows:

```
Routine (Position Control);
Initialize values of Kp, Ti and Td;
Calculate Kd = Kp * Td / T;
Calculate Ki = Kp * T /Ti;
Initialize initial error to zero i.e. d = ei = 0;
Repeat
          Read ADC Output (Ao);
          Calculate error e = setpoint – Ao;
          Calculate integral coefficient d = d + e;
          Calculate output y = Kp * e + Ki * d + Kd (e – ei);
          Update error ei = e;
          Output y to DAC;
```

The top level, i.e. behavioral model was developed in VHDL. Assembly code was developed separately using the picoblaze IDE. The same was executed using KCPSM3.exe. The KCPSM3 VHDL file, generated ROM file are added to the project environment of Xilinx webpack and dumped to Spartan III FPGA using JTAG programmer. The system was tested and found to worked satisfactorily for several trials.

7.2.5 Final SoC Specifications

7.2.5.1 Results and Conclusion

The RTL hierarchical version of the implementation is shown in Fig. 7.2. The Xilinx mapping report file is given below:

Table 7.1 Xilinx mapping report file for design 'temperature controller'

```
Design Information
------------------
temp_controller.pcf
Target Device  : x3s400
Target Package : pq208
Target Speed   : -4
Mapper Version : spartan3 -- $Revision: 1.16 $
```

Fig. 7.4 Block Sche-
matic of the system

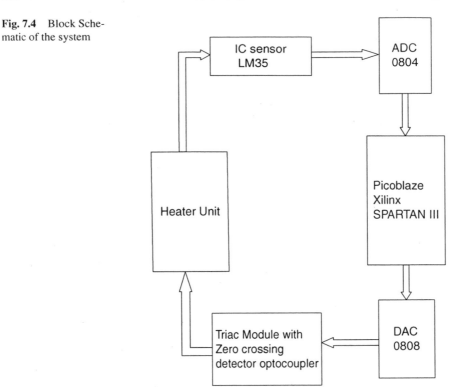

Table 7.2   Design summary reported by Xilinx Webpack

Logic Utilization:
        Number of Slice Flip Flops: 70 out of 7,168   1%
        Number of 4 input LUTs: 109 out of   7,168   1%
Logic Distribution:
        Number of occupied Slices: 97 out of   3,584   2%
        Number of Slices containing only related logic: 97 out of   97
100%
        Number of Slices containing unrelated logic: 0 out of   97   0%

Technically useful portion of the design summary is reproduced below:

It is evident from the above table that the design can be accommodated only in the 1% of the total FPGA resources. The number of occupied slices are 97 occupied completely by the design. This shows that unlike the implementation on traditional microcontrollers such as MCS-51 series or PIC, the implemented design has no room for redundant logic.

Table 7.3   Delay summary report

The NUMBER OF SIGNALS NOT COMPLETELY ROUTED for this design is: 0
The AVERAGE CONNECTION DELAY for this design is: 1.084
The MAXIMUM PIN DELAY IS: 3.560
The AVERAGE CONNECTION DELAY on the 10 WORST NETS is: 2.935

The delay report indicates sound implementation of the temperature controller in the SPARTAN III FPGA with soft core of the Picoblaze. The design offers a rapid prototyping environment to realize the semicustom design and provides a full proof means to evaluate would be implementation of dedicated full custom ASIC.

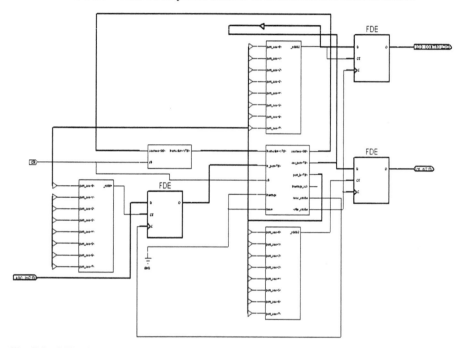

Fig. 7.5   RTL hierarchical version of the implementation

Program 7.3: VHDL code for position algorithm
****
-- Company: Shivaji University
-- Engineer: Shinde S.A and Dr.R.K.Kamat
-- Create Date: 22:45:04 05/30/2008
-- Design Name:
-- Module Name: Position Algorithm - Behavioral
-- Project Name: Picoblaze
-- Target Devices: Spartan 3E 500

```vhdl
-- Tool versions: Webpack 9.2i
-- Description:
-- Revision:
-- Revision 0.01 - File Created

library IEEE;
use IEEE.STD_LOGIC_1164.ALL;
use IEEE.STD_LOGIC_ARITH.ALL;
use IEEE.STD_LOGIC_UNSIGNED.ALL;
---- Uncomment the following library declaration if instantiating
---- any Xilinx primitives in this code.
--library UNISIM;
--use UNISIM.VComponents.all;
entity PositionAlgorithm is
Port (ADC_in: in std_logic_vector(7 downto 0);
                DAC_Out : inout std_logic_vector(7 downto 0);
      clk : in std_logic);
end PositionAlgorithm;
architecture Behavioral of PositionAlgorithm is
component kcpsm3
port (
address : out std_logic_vector(9 downto 0);
instruction : in std_logic_vector(17 downto 0);
port_id: out std_logic_vector(7 downto 0);
write_strobe: out std_logic;
out_port: out std_logic_vector(7 downto 0);
read_strobe: out std_logic;
in_port: in std_logic_vector(7 downto 0);
interrupt : in std_logic;
interrupt_ack: out std_logic;
reset : in std_logic;
clk: in std_logic
);
end component;
component control
port (
address : in std_logic_vector(9 downto 0);
instruction : out std_logic_vector(17 downto 0);
clk: in std_logic
);
end component;
--Processor signals
signal add_bus: std_logic_vector(9 downto 0);
signal inst_bus: std_logic_vector(17 downto 0);
```

```vhdl
signal port_add: std_logic_vector(7 downto 0);
signal data_out: std_logic_vector(7 downto 0);
signal data_in: std_logic_vector(7 downto 0);
signal write1 : std_logic;
signal read1 : std_logic;
begin
processor: kcpsm3
port map(
address => add_bus,
instruction => inst_bus,
port_id=> port_add,
write_strobe=> write1,
out_port=> data_out,
read_strobe=> read1,
in_port=> data_in,
interrupt => '0',
interrupt_ack=> open,
reset => '0',
clk=> clk
);
program: control
port map(
address => add_bus,
instruction => inst_bus,
clk=> clk
);
ADC_Register: process(read1, data_in, port_add)
begin
if port_add= x"00" then
if read1'event and read1 = '1' then
data_in<=ADC_in;
end if;
end if;
end process ADC_register;
DAC_Register: process(write1, data_in, port_add)
begin
if port_add= x"08" then
if write1'event and write1 = '1' then
DAC_Out<=data_out;
end if;
end if;
end process DAC_Register;
end Behavioral;
```
****

Program 7.4: Assembly code for position algorithm
****

```
        adc DSIN 00
        DAC DSOUT 08
        Kp EQU sa
        Ti EQU sb
        Td EQU sc
        T EQU 30
        dividend equ s5
        divisor equ s6
        result equ s7
        multiplier equ s8
        multiplicant equ s9
        temp equ s4
        Kd equ s3
        KI equ s2
        ei equ s0
        setpoint equ 45
        e equ 0
        d equ 0
        value equ sd
        value1 equ se
        value2 equ sf
back: in s0,adc
   call algorithm
   call output
        call delay
        jump back
algorithm: LOADkp, 20
   LOAD        Ti, 25
   LOAD        Td, 28
   ;division of T and Ti
   LOAD dividend, T
   LOAD divisor,Ti
   CALL div
   LOAD        multiplier,Kp
   LOAD        multiplicant,result
   CALL        multiplication
   LOAD        Kd,result
   LOAD        dividend,Ti
   LOAD        divisor,T
   CALL div
   LOAD multiplicant,result
   LOAD multiplier,Kp
   CALL multiplication
```

```
LOAD Ki,multiplier
LOAD ei,0
LOAD s0,adc
load s9,setpoint
SUB s9,s0;      value of e
ADD ei,s9;      value of d e+d
CALL output
output: LOAD multiplier,Kp
    LOAD multiplicant,e
    CALL multiplication
    LOAD value,multiplier
    LOAD multiplier,Ki
    LOAD multiplicant,d
    call multiplication
    LOAD value1,multiplier
    LOAD s1,e
    LOAD s2,ei
    sub s1,s2
    LOAD multiplier,Kd
    LOAD multiplicant,s2
    call multiplication
    LOAD value2,multiplier
    ADD value,value1
    ADD value2,value
    OUT value2,DAC
    load ei,s2
    ret
delay: ret
div:  SUB    dividend, divisor;          do subtraction
      JUMP   C, stop;         jump when carry
      add    result, 1;       otherwise, add 1 to result
      JUMP   div              continue subtraction
stop: RET
multiplication: load temp, multiplier
    sub multiplicant,1
    jump Nz, mul
    load result,multiplier
    ret
mul:  add multiplier,temp
    load result,multiplier
    sub multiplicant,1
    jump NZ, mul
    ret
```

****

References

1. http://www.st.com/stonline/products/technologies/soc/soc.htm
2. http://whatis.techtarget.com/definition/0,,sid9_gci859459,00.html#
3. http://www.answers.com/topic/system-on-a-chip?cat=technology
4. A Survey of Three System-on-Chip Buses:, AMBA, CoreConnect and Wishbone, Milica Mitić and Mile Stojčev, es.elfak.ni.ac.yu/Papers/ICEST%20'06.pdf
5. L. Bennini, G. DeMicheli, Networks on Chips: A New SoC Paradigm, IEEE Computer, Vol. 35, No. 1, January 2002 pp. 70–78.
6. S. Yoshikazu, S. Hiroaki, M. Kouji, T. Satoshi, Present Status of the Embedded CPU in SoC Design, NEC Technical Journal, Vol. 1, No. 5; pp. 38–41, 2006 www.nec.co.jp/techrep/en/journal/g06/n05/t060510.pdf
7. Intel Starts Foray into SoC Market, Christoph Hammerschmidt, EE Times Europe,(07/24/2008 4:11 AM EDT), http://www.eetimes.com/rss/showArticle.jhtml?articleID=209600265&cid=RSSfeed_eetimes_newsRSS
8. Multi-processor SoC design poses debug challenge, by N. Flaherty, Friday 12 January 2007 http://www.electronicsweekly.com/Articles/2007/01/12/40512/multi-processor-soc-design-poses-debug-challenge.htm
9. Embedded Systems White Papers, System-on-Chip Designs: Strategy for Success http://whitepapers.silicon.com/0,39024759,60022283p,00.htm
10. http://www.us.design-reuse.com/articles/2502/system-on-chip-market-to-hit-1-3-billion-units-in-2004-says-new-report.html
11. System on Chip – A Global Strategic Business Report, Research Report # GIA-MCP1471, Publication Date: May 2008 Global Industry Analysts, Number of Pages: 1198
12. SoC market to exceed $58bn by 2010, Global system-on-a-chip or SoC market likely to increase by over 150pc between 2006 and 2010, http://www.ciol.com/content/780798852.aspx
13. Worldwide System-on-Chip (SoC) Market To Reach $43.2 Billion By 2009, By Electronics.ca Research Network Published 01/17/2005 Semiconductors, http://www.electronics.ca/presscenter/articles/48/1/Worldwide-System-on-Chip-SOC-Market-To-Reach-432-Billion-By-2009/Page1.html
14. Mixed Signal System-on-Chip Applications- Cause of Mutual Interest! Date Published: 22 Nov 2005 By C. Vasanthalakshmi, Research Trainee, http://www.frost.com/prod/servlet/market-insight-top.pag?docid=53870501 Frost and Sullivan Report, 2005.
15. Support for On-Chip Memory in Fiasco, Daniel Molka, TU Dresden, Verteidigung der Beleg-Arbeit, Retrieved from http://os.inf.tu-dresden.de/EZAG//abstracts/abstract_20070817.xml
16. System–on-Chip (SoC), Saturday, 06 January 2007 http://www.electronics-manufacturers.com/info/circuits-and-processors/system-on-chip-soc.html
17. SERC: 6th May 2008: Colloquium: "On-Chip Memory Architecture Exploration of Embedded Systems" https://www.serc.iisc.ernet.in/broadcast_messages/msg12231.html

18. System-on-Chip Bus, The next generation of System-on-Chip, Daniel Wiklund, Electronic devices, Department of Physics and Measurement Technology, Linköping University, SE-581 83 Linköping, Sweden

19. Lecture 2a, Overview of System-on-Chip Design, Res Saleh, University of British Columbia, Dept. of ECE

20. SoC bus war fizzles, Ron Wilson, EEdesign.com, (07/01/2002 6:13 PM EDT), http://www.eetimes.com/news/design/columns/design_future/showArticle.jhtml?articleID=17407890

21. FPGA Implementation of DLX Microprocessor With WISHBONE SoC Bus, By R. Selvakumar Rajagopal, M, Mun'im Ahmad Zabidi, Universiti Teknologi Malaysia (UTM), http://www.design-reuse.com/articles/18600/dlx-microprocessor-wishbone-bus.html

22. T.M. Reeves, T.K. Ravey, K. Timothy, Preface- System-on-a-Chip and Packaging, IBM Journal of Research and Development, Vol. 46, No. 6, 2002

23. Physical Limits to Modularity, By Daniel E Whitney, Senior Lecturer MIT Engineering Systems Division, 3/3/2004.

24. The Development of the C Language* Dennis M. Ritchie, Bell Labs/Lucent Technologies, http://cm.bell-labs.com/cm/cs/who/dmr/chist.html

25. Investigation of Hardware JPEG Encoder Implementation and Verification Methodologies, S. Gordoni, Department of Electrical and Computer Engineering, University of California Santa Barbara

26. The four Rs of efficient system design By Juergen Jaeger and Shawn McCloud, Courtesy of Embedded Systems Programming Mar 1 2005 (14:39 PM) URL: http://www.us.design-reuse.com/showArticle.jhtml;jsessionid=NTXNPH3R3BVB4QSND BCSKH0CJUMEKJVN?articleID=60404381

27. Presentation by Jamie Lokier, Liverpool University, UK & CERN, Switzerland

28. The Network Interface Bottleneck, 10th IEEE Real Time Conference, Beaune '97, M. Boosten, R.W. Dobinson, B. Martin, CERN & Stan Ackermans Institute, Eindhoven, [RT97]

29. Hardware Compilation Group, Oxford University Computing Laboratory http://www.comlab.ox.ac.uk/oucl/hwcomp.html, [OUCL]

30. Embedded Solutions Ltd. – Commercial suppliers and support for Handel-C, http://www.embeddedsol.com/, [ESL]

31. C level design http://www.cleveldesign.com/, [ALT]

32. Frontier Design AIRT http://www.frontierd.com/, Handel-C

33. http://www.embeddedsol.com/, NLC

34. Systolic Parallel C http://www-mp.informatik.unimannheim.de/groups/mass_par_1/projects/spc.html

35. Transmogrifier C http://www.eecg.toronto.edu/EECG/RESEARCH/tmcc/tmcc/

36. Handel-C for Hardware Design the Value of Software Design and Debug Methods To Designers Addressing Reconfigurable Logic August 2002White paper by Celoxica

37. Hardware C – A Language for Hardware Design, Descriptive Note: Technical Report, Stanford University CA Computer Systems Lab, Ku, David C.; De Micheli, Giovanni, Aug 1988

38. High level languages- C-FPGA environment, http://asic-soc.blogspot.com/2007/11/advanced-tools-in-reconfigurable.html

39. What is Transmogrifier C ? http://www.eecg.toronto.edu/RESEARCH/tmcc/tmcc/

40. C Level Design Introduces C2Verilog Version 2.0 to Increase Productivity for Electronic System Designers, Business Wire, Dec 15, 1998, http://findarticles.com/p/articles/mi_m0EIN/is_1998_Dec_15/ai_53400612

41. C to FPGA: An Abstract Concept for Concrete Design Implementation, By Jeff Jussel, Celoxica, http://www.rtcmagazine.com/home/article.php?id=100304

42. Spec-C, Handel-C, SystemC: A Comparative Study, By: Nikola Rank, 13 March 2006 www.web.cecs.pdx.edu/~mperkows/CAPSTONES/DSP1/ELG6163_Rank.ppt

43. System-on-chip design methodology in engineering education by William D. Mensch, Jr.1 and Dennis A. Silage2 retrieved from www.temple.edu/scdc/icee2000.pdf

44. Processor cores put more into SoC designs, retrieved from http://www.electronicstalk.com/news/tey/tey174.html

45. Programmable Clock Generator Solves System-Timing Woes, Dave Bursky I ED Online ID #1720 I December 23, 2002, http://electronicdesign.com/Articles/Index.cfm?AD=1&ArticleID=1720

46. A SoC for Multimedia Network Devices, T. Boesch, E. Roth, M. Thalmann, N. Felber, W. Fichtner, ftp://ftp.tik.ee.ethz.ch/pub/people/spin/icce.pdf

47. SoC Design Environment with Automated BusArchitecture Generation for Rapid Prototyping with ISS, Sang-Heon Lee1, Jae-Gon Lee1, Ando Ki2, Chong-Min Kyung1, vs. www.kaist.ac.kr/~kyung/html/Paper/International%20Conference/IC-(132).pdf

48. Antifuse Memory IP Fuels Low-Power Designs, Jim Lipman,(08/01/2008 6:00 PM EDT), URL: http://www.eetimes.com/showArticle.jhtml?articleID=209900470

49. Lower voltage next goal for low-power DDR, Marc Greenberg,(06/09/2008 12:00 AM EDT), URL: http://www.eetimes.com/showArticle.jhtml?articleID=208402252

50. Keeping the best audio quality in mobile phone by managing voltage drops created by 217 Hz transients, D & R Industry Articles, http://www.design-reuse.com/articles/18519/managing-voltage-drops.html

51. Enhance circuit timing design with programmable clock generators (Part 1 of 2), As clocks speed increase and the number of clocks increases, a programmable clock generator may offer a better system and EMI design solution, By Lin Wu, Product Marketing Manager, Texas Instruments, Planet Analog, (06/03/08, 05:36:04 PM EDT) http://www.design-reuse.com/exit/?url=http%3A%2F%2Fwww.embedded.com%2Fcolumns%2Ftechnicalinsights%2F208402074%3Fprintable%3Dtrue

52. Enhance circuit timing design with programmable clock generators (Part 2 of 2), As clocks speed increase and the number of clocks increases, a programmable clock generator may offer a better system and EMI design solution, By Lin Wu, Product Marketing Manager, Texas Instruments, Planet Analog, (06/04/08, 12:00:00 PM EDT)

53. UWB Time-interleaved ADC exploiting SAR, Silvia Dondi, Silis s.r.l., Marco Bigi, Andrea Boni, Matteo Tonelli, Dipartimento di Ingegneria dell'Informazione – University of Parma, http://www.design-reuse.com/articles/17552/uwb-time-interleaved-adc-sar.html

54. B. Le, T. W. Rondeau, J. H. Reed, C. W. Bostian, "Analog-to-Digital Converters", IEEE Signal Processing Magazine, Nov. 2004

55. Future-Ready Ultrafast 8bit CMOS ADC, for System-on-Chip Applications, Jincheol Yoo1, Daegyu Lee1, Kyusun Choi1, and Ali Tangel2, www.cse.psu.edu/~kyusun/res/wasic.pdf

56. T. Monnier, F.M. Roche, G. Cathebras, Flip-flop hardening for space applications, Memory Technology, Design and Testing,1998. Proceedings. International Workshop on Volume, 24–25 pp. 104–107, Aug 1998

57. US Patent 6501314 – Programmable differential D flip-flop, retrieved from http://www.patentstorm.us/patents/6501314/description.html

58. I. Shigeru, S. Masaaki, N. Sen, K. Akio, A Realization of Asymmetrically Faulty D-flip/flop Based on Power Flicking Reset and Its Application to Fail-safe systems, IEIC Technical Report, Institute of Electronics, Information and Communication Engineers, Vol. 101; No.3(FTS2001 1 -13);pp. 49–55, 2001

59. Phase Frequency Detector with a Novel D Flip Flop, Kaben Research Inc, retrieved from http://www.wipo.int/pctdb/en/wo.jsp?wo=2005096501

60. Flip-flop (electronics), http://en.wikipedia.org/wiki/Flip-flop_(electronics)

61. Fundamentals of the Electronic Counters, Application Note 200, Electronic Counter Series, Agilent Technologies, Electronic Counter Measures, Authored by Lee Pucker, Spectrum Signal processing

62. Introduction to CMOS VLSI Design, SRAM, Lecture Notes Retrieved from www.fp.cse.wustl.edu/dzar/463/Lectures/lect13.ppt

63. A New Low Power and High Speed Bidirectional Shift Register Architecture, N. Sklavos, P. Kitsos, N. Zervas and O. Koufopavlou, Retrieved from www.patmos2001.eivd.ch/program/Repro%5CArt_10_4.pdf

64. Frequency divider design strategies, By Louis Fan Fei, Broadband Technology, Retrieved from www.rfdesign.com/mag/503rfdf1.pdf

65. Fifo Controller Design Technique in JPEG 2000 Encoder, R.S. Gamad, Hemant Saxena, retrieved from www.hindawi.com/RecentlyAcceptedArticlePDF. aspx?journal=MSE&number=694582

66. VLSI Design and Verification of the Imagine Processor, Brucek Khailany, William J. Dally, Andrew Chang, Ujval J. Kapasi, Jinyung Namkoong, Brian Towles, Proceedings of the 2002 International Conference on Computer Design, Retrieved from www.cva.stanford. edu/publications/2002/khailany_iccd2002_imagine_impl.pdf

67. Embedded RAM In FPGAs Enables FIFO Applications Creating A FIFO Memory Solves The Problem Of An Asynchronous Boundary Between Clocks, ED Online ID #7522, March 8,1999, Retrieved from http://electronicdesign.com/Articles/Index. cfm?ArticleID=7522&pg=1

68. Using DMA FIFO to Develop High-Speed Data Acquistion Applications for Reconfigurable I/O Devices, NI Application Note, Retrieved from http://zone. ni.com/devzone/cda/tut/p/id/4534

69. FIFO memories supply the glue for high-speed systems, Markus Levy, EDN,march 1997 Retrieved from http://www.edn.com/archives/1997/031497/06DF_02.htm

70. P. Balasubramanian, K. Anantha, Power and Delay Optimized Graph Representation for Combinational Logic Circuits, International Journal of Computer Science Vol. 2 Retrieved from www.waset.org/ijecs/v1/v1-1-2.pdf

71. P. Balasubramanian, C. Hari Narayanan, K. Anantha, Low Power Design of Digital Combinatorial Circuits with Complementary CMOS Logic, International Journal of Electronics, Circuits and Systems Vol. 1 No. 1

72. E. Bareiša, V. Jusas, K. Motiejūnas, R. Šeinauskas, Delay Fault Models and Metrics, Information Technology and Control, Vol. 34, No. 4, 2005, ISSN 1392 – 124X

73. Soft Error Resilient System Design through Error Correction, Subhasish Mitra, Ming Zhang, Norbert Seifert, TM Mak, Kee Sup Kim, Retrieved from www.gigascale.org/ pubs/893/ifip06%5B1%5D.final.v7.pdf

74. J.G. Delgado-Frias, J. Nyathi, A VLSI high-performance encoder with priority lookahead, VLSI,1998. Proceedings of the 8th Great Lakes Symposium on VLSI design, Vol. 19–21, pp. 59–64, Feb 1998

75. IP reuse simplifies SoC design, verification John Wilson, EE Times,(10/11/2004), http:// www.eetimes.com/news/latest/showArticle.jhtml?articleID=49900412

76. Design Reuse, http://whatis.techtarget.com/definition/0,sid9_gci759468,00.html

77. R. Mathur, Aptix Corporation, SoC Prototyping Requirements, FPGA and Structured ASIC Journal, Retrieved from http://www.fpgajournal.com/articles/soc_aptix.htm

78. C- Based Rapid Prototyping for Digital Signal Processing, E. Casseau, B. Le gal, P. Bomel, C. Jego, S.Huet, E. Martin, Retrieved from www.eurasip.org/Proceedings/Eusipco/ Eusipco2005/defevent/papers/cr1179.pdf

79. Teaching IP Core Development: An Example, Aleksandar Milenkovic, David Fatzer, Retrieved from http://www.ece.uah.edu/~milenka

80. ASIC Prototyping Using Off-the-Shelf, FPGA Boards: How to Save Months of Verification Time and Tens of Thousands of Dollars, Retrieved from www.synplicity.com/literature/ whitepapers/pdf/proto_wp06.pdf

81. L.A. Zadeh, Fuzzy Logic, IEEE Computer, pp. 83–89, 1988

82. L.C. Jain and N.M. Martin, Fusion of Neural Networks, Fuzzy Sets, and Genetic Algorithms: Industrial Applications, The CRC Press, 1999

83. A. Kandel, G. Langholz, Fuzzy Hardware: Architectures and Applications, Springer, 1998.

84. Handel C reference manual www.celoxica.com/techlib/files/CEL-W0410251JJ4-60.pdf

85. V. Paxson, K. Asanovic, S. Dharmapurikar, J. Lockwood, R. Pang, R. Sommer, N. Weaver: Rethinking Hardware Support for Network Analysis and Intrusion Prevention, USENIX First Workshop on Hot Topics in Security (HotSec), Vancouver, B.C., July 31, 2006

86. S. Sachidananda, S. Gopalan, S. Varadarajan, Hardware-Software Hybrid Packet Process-
     ing for Intrusion Detection Systems, Vol. 3802/2005, Springer, 2005
87. D.V. Schuehler, J.W. Lockwood, TCP Splitter. A TCP/IP Flow Monitor in Reconfigurable
     Hardware, IEEE Micro, Vol. 23, No. 1, pp. 54–59, Jan/Feb 2003
88. D. Bertozzi, A. Jalabert, S. Murali, R. Tamhankar, S. Stergiou, L. Benini, G. De Micheli:
     NoC Synthesis Flow for Customized Domain Specific Multiprocessor Systems-on-Chip,
     2005
89. E. Bolotin, I. Cidon, R. Ginosar, A. Kolodny: QNoC: QoS Architecture and Design Process
     for Networks on Chip, JSA, Feb 2004.
90. K. Goossens, J. Dielissen, A. Radulescu: A Ethereal Network on Chip: Concepts, Architec-
     tures, and Implementations, IEEE Design and Test of Computers, Sep/Oct, 2005.
91. F. Moraes, N. Calazans, A. Mello, L. Möller, L. Ost, Hermes: an Infrastructure for Low
     Area Overhead Packetswitching Networks on Chip, Integration, VLSI Journal, Oct. 2004.
92. R. Gindin, I. Cidon, I. Keidar, NoC-Based FPGA: Architecture and Routing, http://www.
     ee.technion.ac.il/matrics/papers/NoC-Based%20FPGA.pdf Retrieved on March 1, 2008
93. D.H. Lehmer. Mathematical methods in large-scale computing units. In Proc. 2nd Sympos.
     on Large-Scale Digital Calculating Machinery, Cambridge, MA, 1949, pp. 141–146, Cam-
     bridge, MA, 1951. Harvard University Press.
94. Application-specific integrated circuit, http://www.answers.com/topic/asic
95. ALD Full Custom Design, Retrieved from http://www.aldinc.com/ald_fullcustom.htm
96. W.J. Dally, A. Chang, The Role of Custom Design in ASIC Chips, Retrieved from www.
     cva.stanford.edu/publications/2000/dac00.pdf
97. B. Kick, U. Baur, J. Koehl, T. Ludwig, T. Pflueger, Standard-cell-based design methodology
     for high-performance support chips, IBM Journal of Research and Development, Jul-Sep
     1997
98. Under the Hood: Gauging standard-cell performance, by Michael Keller, Semiconduc-
     tor Insights 08/27/2007, Retrieved from URL: http://www.eetimes.com/showArticle.
     jhtml?articleID=201801414
99. N. Sherwani. Algorithms For VLSI Physical Design Automation. 3rd ed., Kluwer Academic
     Publishers, pp. 222–223, 1999.
100. System-on-Chip Designs Strategy for Success, W H I T E P A P E R – June 2001 Udaya
     Kamath Rajita Kaundin, Retrieved from www.wipro.com/pdf_files/Wipro_System_on_
     Chip.pdf
101. Where Do Structured ASICs Fit?, Retrieved from http://www.soccentral.com/results.
     asp?CatID=468
102. Assessing the structured-ASIC alternative, Ron Wilson, EE Times, 06/03/2004,
     Retrieved   from   http://www.eetimes.com/industrychallenges/silicon/showArticle.
     jhtml?articleID=21401237
103. Embedded FPGA soft core processor enables universal CompactPCI applications, Pat
     Mead, Marketing Manager Altera Europe And Barbara Schmitz, Marketing Director MEN
     Mikro Elektronik, Embedded System Engineering Magazine, Retrieved from http://www.
     esemagazine.com/index.php?option=com_content&task=view&id=187&Itemid=2
104. Instrumentation control using the Rabbit 2000 embedded microcontroller, Authors:
     Schofield, Ian S.; Naylor, David A., Advanced Software, Control, and Communication Sys-
     tems for Astronomy. Edited by Lewis, Hilton; Raffi, Gianni. Proceedings of the SPIE, Vol.
     5496, pp.dr 392–401 (2004)
105. FPGA based CPU instrumentation for hard real-time embedded system testing, Richard
     Fryer, ACM SIGBED Review archive, Vol. 2, No. 2 (April 2005) table of contents, Special
     issue: IEEE RTAS 2005 work-in-progress, pp. 39–42, 2005, ISSN:1551–3688
106. Exploring C for Microcontrollers: A Hands on Approach by J. S. Parab, V. G. Shelake, R.
     K. Kamat and G.M. Naik Springer Netherlands, 2007
107. Altera tweaks soft-core processor, By Susan Rambo, Embedded.com, 02/25/03, http://
     www.embedded.com/story/OEG20030225S0022 retrieved on October 20, 2007

108. QuickLogic QuickMIPS ESP Is Industry's First – Combines a High-speed Processor With Hardwired Functions and Field Programmability, Design and Reuse Newsletter, http:// www.us.design-reuse.com/news/news349.html retrieved on October 21, 2007

109. PicoBlaze User Resources Xilinx Inc. website http://www.xilinx.com/ipcenter/processor_ central/picoblaze/picoblaze_user_resources.htm retrieved on October 21, 2007

110. "Proportional-integral-derivative explained, Tuning PID controls", Javier Gutirrez, National Instruments, Industrial Control Design Line, CMP Media, http://www.industrialcontrold-esignline.com/199000752; jsessionid=XZELCWV0SLJ2QQSNDLQCKH0CJUNN2JVN? printableArticle=true retrieved on October 20, 2007

111. "New techniques for PID controller design", Moradi, M.H., Control Applications, 2003. CCA 2003. Proceedings of 2003 IEEE Conference on Vol. 2, No. 23–25 June 2003pp: 903–908 vol.2

112. The unified gain tuning approach to the PID position control with minimal overshoot, position stiffness, and robustness to load variance for linear machine drives in machine tool environment Joohn-Sheok Kim Moon-Suk Choi Seokjoo Kan, Dept. of Electr. Eng., Inchon Univ., Applied Power Electronics Conference and Exposition, 2001. APEC 2001. Sixteenth Annual IEEE, Publication Date: 2001, Vol. 1, On page(s): 635–641 Vol. 1, ISBN: 0-7803-6618-2

113. The PID Algorithm, STRAIGHT-LINE CONTROL CO., INC., http://members.aol.com/ pidcontrol/pid_algorithm.html

114. Xilinx Reference Design License http://www.xilinx.com/ipcenter/doc/referencedesignli-cense.pdf

115. PicoBlaze Soft Processor homepage http://www.xilinx.com/ipcenter/processor_central/ picoblazer/index.htm

116. KCPSM3 Manual: www.xilinx.com/bvdocs/userguides/ug129.pdf

117. Lecture notes Digital Control Algorithms and Their Implementation by Peyman Gohari http://users.encs.concordia.ca/~gohari/ELEC6061/Files/9.pdf

118. DATA ENCRYPTION STANDARD, U.S. DEPARTMENT OF COMMERCE/National Institute of Standards and Technology, FEDERAL INFORMATION PROCESSING STANDARDS PUBLICATION, FIPS PUB 46-3, 1999 October 25, Retrieved from http:// csrc.nist.gov/publications/fips/fips46-3/fips46-3.pdf on July 1, 2008.

119. Garfinkel, Simson,(December 1, 1994). PGP: Pretty Good Privacy. O'Reilly Media, pp.101–102. ISBN 978-1565920989.

120. Lin, M.C.-J. Youn-Long Lin, A VLSI implementation of the Blowfish encryption/decryp-tion algorithm, Design Automation Conference,2000. Proceedings of the ASP-DAC 2000. Asia and South Pacific

121. Cody, Brian; Madigan, Justin; MacDonald, Spencer; Hsu, Kenneth W, High speed SOC design for blowfish cryptographic algorithm, Very Large Scale Integration, 2007. VLSI - SoC 2007. IFIP International Conference, 2007

122. FPGA to ASIC Strategy for Communication SoC Designs, by Rick Mosher, AMI Semicon-ductor, Retrieved from http://www.design-reuse.com/articles/4360/fpga-to-asic-strategy-for-communication-soc-designs.html

123. FPGA-based System-on-Chip Designs for Real-Time Applications in Particle Physics, Shebli Anvar, Olivier Gachelin, Pierre Kestener, Herve Le Provost, Irakli Mandjavidze, Presented at the 14th IEEE Real Time Conference, Stockholm, Sweden, June 6–10, 2005, Retrieved   from   www.irfu.cea.fr/Phocea/file.php?class=std&&file=Doc/Publications/ Archives/dapnia-05-105.pdf

124. The Half-Adder Form and Early Branch Condition Resolution, David R. Lutz and D. N. Jayasimha, Proceedings of the 13th Symposium on Computer Arithmetic (ARITH '97)

125. B. Adrian, T. Jarmo, Vlsi-efficient implementation of full adder-based median filter,2004 IEEE International Symposium on Circuits and Systems: Proceedings, May 23–26, 2004, Sheraton Vancouver Wall Centre Hotel, Vancouver, British Columbia, Canada

126. D.J. Soudris, V. Paliouras, T. Stouraitis, C.E. Goutis, VLSI design methodology for RNS full adder-based inner product architectures, IEEE transactions on circuits and systems. 2, Analog and digital signal processing ISSN 1057–7130

127. Adders and computational power, Retrieved from http://www.edacafe. com/books/phdThesis/Chapter-2.5.php

128. Booth Recoding, ASIC Design for Signal Processing, Retrieved from http://www. geoffknagge.com/fyp/booth.shtml

129. IP & Ethernet Interfaces, Retrieved from http://www.beyondlogic.org/etherip/ip.htm

130. S. A. Shinde, V. G. Shelake, R. K. Kamat, FPGA based Packet Splitter Implementation Using Mixed Design Flow, ELECTRONICS AND ELECTRICAL ENGINEERING, 2008. No. 8(88)

Index

Printed in the United States
142974LV00003B/55/P

9 781402 093616